ETHICS AND BIOTECHNOLOGY

SOCIAL ETHICS AND POLICY
Edited by Anthony Dyson and John Harris
Centre for Social Ethics and Policy,
University of Manchester

EXPERIMENTS ON EMBRYOS
Edited by Anthony Dyson and John Harris

THE LIMITS OF MEDICAL PATERNALISM
Heta Häyry

PROTECTING THE VULNERABLE
Autonomy and Consent in Health Care
Edited by Margaret Brazier and Mary Lobjoit

MEDICAL CONFIDENTIALITY AND LEGAL PRIVILEGE
Jean V. McHale

ETHICS AND BIOTECHNOLOGY

Edited by
Anthony Dyson and John Harris

London and New York

First published 1994
by Routledge
11 New Fetter Lane, London EC4P 4EE

Simultaneously published in the USA and Canada
by Routledge
29 West 35th Street, New York, NY 10001
Reprinted 1998, 2000

Routledge is an imprint of the Taylor & Francis Group

Typeset in 10/12 pt Garamond by
Florencetype Ltd, Kewstoke, Avon
Printed and bound in Great Britain by
T. J. I. Digital, Padstow, Cornwall.

Printed on acid free paper

British Library Cataloguing in Publication Data
A catalogue record for this book is available from the British Library

Library of Congress Cataloging in Publication Data
Ethics and biotechnology / [edited by] Anthony Dyson and John Harris.
p. cm. – (Social ethics and policy)
Includes bibliographical references.
1. Biotechnology – Moral and ethical aspects. I. Dyson, Anthony
II. Harris, John. III. Series.
TP248.2.E85 1993
174′.9574–dc20
93–15482

ISBN 0–415–091403

CONTENTS

FIGURES

NOTES ON CONTRIBUTORS

Stephen R. L. Clark is Professor of Philosophy at Liverpool University. His published work includes *The Moral Status of Animals* (1977), *The Nature of the Beast* (1982), *Civil Peace and Sacred Order* (1989) and *How to think about the Earth* (1993). He is co-editor of the *Journal of Applied Philosophy*.

David Colman is Professor of Agricultural Economics at Manchester University and is in charge of the national Farm Business Survey in the north-west of England. He has written extensively on agricultural policy, commodity markets and the economics of less-developed countries. His publications include (with F. Nixon) *Economics of Change in Less-Developed Countries* (Philip Allan, 1986 2nd edn) and (with T. Young) *Principles of Agricultural Economics* (Cambridge University Press, 1988).

Søren Holm is Research Fellow at the Institute of Biostatistics and Theory of Medicine, University of Copenhagen. He has qualifications in medicine and health-care ethics.

Hillel Steiner is Senior Lecturer in Political Philosophy at the University of Manchester. A member of the editorial boards of *Ethics* and *Social Philosophy & Policy* and of the advisory board of the Centre for Social Ethics and Policy, he has published widely on the subjects of justice, liberty and moral reasoning and is the author of *An Essay on Rights* (forthcoming).

Bonnie Steinbock is Professor and Chair of the Department of Philosophy at the University at Albany. She has published articles in applied ethics and biomedical ethics. She edited *Killing and Letting Die* (1980), and is the author of *Life before Birth: The Moral and Legal Status of Embryos and Fetuses* (Oxford University Press, 1992).

Susan Kimber is a Senior Lecturer in the School of Biological Sciences at the University of Manchester. She has carried out extensive research on early embryonic development and reproductive biology and has had close associations with IVF clinics including working as an adviser.

Janice Wood-Harper is a Lecturer in Biology and Health Studies at Salford College. She is a recent graduate in Health-Care Ethics from the Centre for Social Ethics and Policy, University of Manchester, and her current research interest is the ethics of genetic testing.

Heta Häyry is Docent and Assistant Professor of Practical Philosophy at the University of Helsinki. She is an adviser to the Finnish National Board of Health and Social Security on social ethics and health care ethics, and a member of the Research Council for the Humanities in the Academy of Finland. Her publications in English include *The Limits of Medical Paternalism* (1991) and many articles on health-care ethics.

Charles A. Erin holds Master's Degrees in Nuclear and Elementary Particle Physics and in Technical Change and Industrial Strategy and is Research Fellow at the Centre for Social Ethics and Policy, University of Manchester, for the Commission of the European Communities' Biomedical and Health Research Programme Project 'AIDS: Ethics, Justice and European Policy'. He has published on the ethics of the new reproductive technologies.

Peter R. Wheale is a Director of Bio-Information (International) Limited and Chairperson of the Biotechnology Business Research Group. He has a Ph.D. in Economics, an MA in Medical Ethics and Law and an M.Sc. in the Structure and Organization of Science and Technology. He is co-author of *People, Science and Technology* 1986; *Genetic Engineering: Catastrophe or Utopia* (1988) and co-editor of *The Bio-Revolution: Cornucopia or Pandora's Box?* (1990) and *Animal Genetics: Of Pigs, Oncomice and Men* (1993).

Ruth M. McNally is a Director of Bio-Information (International) Limited and is currently acting International Liaison Officer in the Research Bureau of Brunel University. She has a degree in Genetics and is co-author of *Genetic Engineering: Catastrophe or Utopia* (1988) and co-editor of *The Bio-Revolution: Cornucopia or*

Pandora's Box? (1990) and *Animal Genetics: Of Pigs, Oncomice and Men* (1993).

Matti Häyry is Docent and Junior Research Fellow in Practical Philosophy at the University of Helsinki and Docent of Bioethics at the University of Tampere. He is also an adviser to the Finnish National Board of Health and Social Security on bioethics. His publications in English include *Critical Studies in Philosophical Medical Ethics* (1990) and articles on bioethics in academic journals.

John Harris is Research Director of the Centre for Social Ethics and Policy and Professor of Applied Philosophy in the University of Manchester. Among his many publications on bioethics and applied ethics more generally is *Wonderwoman and Superman: The Ethics of Human Biotechnology* (Oxford University Press, 1992).

Ulla Wessels is Assistant Professor at the University of the Saarland, Saarbrücken, Germany. She has published articles both on applied and theoretical ethics; books she has co-edited include *ANALYOMEN – Perspectives in Analytical Philosophy* (1993), *Praktische Rationalität* (1993) and *Preferences* (1994).

Anthony Dyson is Samuel Ferguson Professor of Social and Pastoral Theology, and Academic Director of the Centre for Social Ethics and Policy, at the University of Manchester. He co-edited, with John Harris, *Experiments on Embryos* (1990).

INTRODUCTION

The development of biotechnology has produced nothing short of a revolution in our capacity not only to manipulate living things from single plant cells to human nature itself, but also to manufacture new life forms almost 'from scratch'. This power to shape and create forms of life has sometimes been described as the power to 'play God'. In this sense the power is not radically different from powers which humans have always possessed and exercised. The dilemma of power is not whether to exercise it – we literally can't help ourselves – but how to exercise it.

Some would dispute the claim that we cannot help ourselves, but there is an important sense in which all changes to the world from 'the fall of a sparrow' to the creation of a new life form make an irrevocable difference. Since we cannot avoid making a difference, it is our moral responsibility to decide what difference to make. This book is about the ethics of deciding what difference to make in the field of biotechnology. It is a vast field, ranging from plant breeding, agriculture, animal breeding and experimentation to human genetics and genetic engineering. We have not tried to be comprehensive, but in the essays that comprise this book we hope to have illustrated both the extent of the dilemmas posed by biotechnology and their fascination and importance.

Finally and equally importantly we hope and believe that the essays in this volume go some way to helping in the resolution of the moral dilemmas posed by biotechnology or at least to showing why some of them are so intractable.

In 'Modern errors, ancient virtues', Stephen Clark deals with a very important preamble to human biotechnology. Having noted the way in which biotechnology may eventually be used to tailor-make human individuals to our specification Clark points out that the first 'victims' of human biotechnology will be animals.

Clark's concern is to set before us an account of what our 'right relations with the animals on whom we – and biotechnologists – rely might be'. As Clark rightly perceives, unless we are confident about the rightness of manipulating animals, we will scarcely be able to move on to human biotechnology, because it is only by animal experimentation that we can, in the medium term, test both the techniques and the products which will enable us to manipulate the human genome. It is thus to the issue of animal experimentation that Clark devotes his chapter.

Clark draws interesting parallels between the treatment of animals by scientists and the treatment of animals by most people in their everyday lives and which is embedded in our culture. He notes that 'The radical charge against experimentalists is not that they are unusually cruel or negligent, but that they act out – in a particularly noticeable form – assumptions deeply ingrained in contemporary culture which are actually false or wicked.' It is this charge that scientific practice is based upon both false and wicked assumptions that Clark attempts to sustain.

Clark develops his argument by considering a number of principles or assumptions upon which the relationship between humans and animals has been based. He labels them egoism, humanism, utilitarianism, and objectivism. Having reviewed them and outlined their deficiencies Clark recommends that we return to what he calls 'ancient virtues' because the principles and assumptions that he has criticized, based as they are on what he calls destructive scepticism, have been found wanting.

The ringing conclusion with which he leaves us is an admirable and cautionary reminder of both the power and the responsibility of human biotechnology. 'Our duty', Clark insists, 'is to admire and to sustain the world in beauty, and not to impose on others pains and penalties we could not ourselves bear.'

While Stephen Clark looks at the impact of biotechnology upon the animal world, in 'Biotechnology and agriculture', David Colman, an agriculture economist, considers its impact on world agriculture. He notes wide agreement that the biotechnology revolution will intensify the industrialization of agriculture and that the consequence of this will be that the dependence of less-developed countries upon the high technology of the West will increase. This will have the effect of radically changing traditional methods of farming in both plant and animal breeding.

Colman outlines the principal directions of biotechnological development and charts their likely impact on world agriculture. He notes that the contingent fact that bio-engineered products are capable of meeting the inventive step criterion of patent ability has enabled both micro-organisms and transgenic animals to be patented in the United States and possibly elsewhere in the world. This has lead to an increasing dominance of agriculture by private industry, which again tends to leave power in the hands of a few western-based multinational companies. The sad lesson is that the 'system by which agricultural biotechnology is delivered will favour the "have's" rather than the "have-nots"'.

Colman's warning, and it is one which it may be already too late to heed, is that 'The balance of economic power will switch increasingly to industry, to high-technology large farms and against smaller farmers in disadvantaged regions and countries'. Colman notes with some considerable sadness that this is seemingly an inevitable consequence of what we regard as economic progress.

Søren Holm, in his 'Genetic engineering and the North–South divide', looks at a different aspect of the theme developed by Colman, namely the tendency of biotechnology to create problems of distributional justice on a global scale. While Colman concentrated principally on charting this tendency with respect to agriculture, Holm looks at it as a doctor and a philosopher and focuses more sharply on the question of justification.

Holm notes the possibilities opened up by genetic engineering that in the future we may be able to engineer protection against infectious diseases and environmental pollutants into the human genome. If this becomes possible, its benefit will be more significant in the developing world than in the industrialized world, and yet he notes that it is almost inevitable that 'enhancing genetic engineering is not going to be widely applied where the need is greatest'. Holm produces an interesting comparison between the vaccination rate in the developing world and in the developed countries. He notes that whereas the developed countries have a vaccination rate between 95 per cent and 98 per cent, the undeveloped countries have at best 40–80 per cent and that in the poorest countries the rate barely reaches 25 per cent. He rightly observes that since vaccination will always be relatively easy compared with genetic engineering, the vaccination rates must be taken as the upper limit of the possible penetration of genetic engineering into developing societies.

Holm concludes that the benefits of enhancing genetic engineering that is, genetic engineering that enhances human capacities and powers, may not be as great as they initially appear. However, they will be sufficiently significant to raise crucial questions of distributive justice both between the developed and the developing world and also within developed countries. For this reason Holm believes that enhancing genetic engineering must be strictly regulated from the perspective of justice.

Hillel Steiner, in his essay 'The fruits of body-builders' labour', looks at our genes and the abilities, powers and capacities with which they endow us – our genetic endowments – from the perspective of political theory. He applies Historical Entitlement Theory to show how and to what extent we may be regarded as the owners of our own bodies and our body products, which of course cover not only our organs, tissue and other physical body products, but also our labour, which is also produced by our bodies. He suggests that in order to own ourselves and the products of our bodies we must own the things that contributed to the substance of our bodies; and in so far as we have benefited from this appropriation, we must compensate those who have not benefited equally.

Steiner's argument is thus that, properly understood, Historical Entitlement Theory, applied to genetic endowment, necessitates a radical redistribution of wealth from those with profitable genetic endowments to those with less profitable ones. His essay shows how the discoverers of molecular and cellular biology give new vigour, new bodily substance, to the somewhat antique fulminations of Sir Robert Filmer and others.

Bonnie Steinbock's essay echoes themes developed by Steiner. She too is interested in the status of the human body, and particularly the embryonic human body, as property. However, her concerns are more rooted in jurisprudence than in the history of political thought. Her interests in 'The moral status of extracorporeal embryos: pre-born children property or something else' focus not so much on the products of the human body as property as on the property we may have in the human body itself. She looks at a number of prominent legal cases concerning jurisdiction over embryos that are extracorporeal prior to implantation. Steinbock discusses the question of whose interest should prevail where the custody or control of extracorporeal embryos is in question. She distinguishes custody from 'control': custody implies that the embryos are children with children's rights, whereas control, she believes, bears no implication.

She believes that many of the cases so far decided, at least in the United States, have been used primarily as a vehicle for judges and indeed other parties to vent their prejudices concerning abortion and related issues. She argues rationally that the relevant interests are not those of the embryo but those of the disputing parties, and that unless this is recognized the attempt to clarify and resolve disputes concerning ownership or control of such embryos will be hopeless.

Susan Kimber writes as an embryologist and a Christian. In her 'IVF and manipulating the human embryo' she provides an interesting balance between the demands of Christian theology and ethics and the special knowledge that a working cell biologist brings to the study of human origins. She surveys not only the scientific development of the human embryo but the various forms of manipulation that it does and can undergo, from a Christian perspective. Her essay is an interesting attempt to marry the demands of science and the imperatives of religion. She shows that Christian ethics can accommodate itself to scientific advance and in particular that some manipulation of, and even experimentation upon, the human embryo can be countenanced within an ethics, the foundations of which were laid in pre-Christian and certainly pre-scientific ages.

As we move more towards the future use of biotechnology, Janice Wood Harper in her 'Manipulation of the germ line: towards elimination of major infectious diseases?' provides a detailed and scholarly survey of the various ways in which manipulation of the germ line may in the near and mid-term have both useful and therapeutic applications. She then goes on to examine the ethical arguments for and against such manipulation. Finally she addresses the issue of safeguards and controls.

Her conclusions are liberal, and on the whole she embraces the new technology and the good that it can achieve both for human individuals and society. She is radical where the objectives and the consequences of experimentation are clearly in view, but cautious and conservative with respect to the dilemmas posed for future generations. She argues coherently that respect for such generations and the views that they may come to form must be a predominant concern in any debate on the moral permissibility of a revolutionary medical procedure which will directly affect both.

In 'How to assess the consequences of genetic engineering', Heta Häyry's aim is to sort out and analyse, first, the major consequences of genetic engineering, and second, the most important ideological factors which cause dispute concerning their weight. For the purpose of the essay, Häyry takes as already decided that there are no intrinsic ethical grounds for condemning genetic engineering.

Häyry proceeds to give an account of the advantages and disadvantages as seen by their respective proponents. On the surface this would not seem to be a difficult undertaking since advantages and disadvantages have already been well publicized. But that is to ignore the formative influence which the social and political setting can exercise upon those who are seeking to make up their ethical minds.

The author first considers what are held to be the advantages of biotechnology, namely in its actual and potential contributions to medicine, pharmacy, agriculture, the food industry and the preservation of our natural environment. Häyry then looks at the alleged disadvantages which are closely related to the alleged advantages. One path of enquiry is to assess risks and costs of the various practices. Often it turns out that the genetic practices can have both good and bad consequences.

But it is also clear that the differences of ethical opinion are not always only factual in character. In particular, people are either 'technological voluntarists' or 'technological determinists'. According to the former, control is ultimately in human hands; according to the latter it is not. Häyry recognizes some truth in this distinction. All in all, however, she takes the view that there are grounds for 'moderate optimism concerning the controllability of biotechnology'.

The author turns next to consider 'the three stages of technological development'. First, there is the theoretical invention. Second, theory is transformed into practical innovations. Third, the completed products are marketed and distributed. Because the second stage is often protracted, public interest, which may have been lively at the first stage, often fades away. But it is essential that the ordinary citizen does not become a psychological defeatist, for the defeatism can be self-fulfilling. Häyry is fairly confident that regulation will be an effective safeguard in most respects. She leaves to the end of her essay a reference to the argument that only the wealthy and the privileged in the West will benefit from genetic engineering. In reply, the author argues that the genetic engineering should not be blamed for the injustice in its implementation. With her 'moderately optimistic view' she proposes that our best chance is to believe that this wide and just distribution is possible, and then proceed to make it possible by our own actions.

Cutting across these topics is the issue of the role and relationships of companies and international agencies. Charles Erin in 'Who owns Mo?' presents a case-study of ethical questions of property and ownership in biotechnology, in particular the use of human biological materials to produce cultured cell lines which may have many biological applications and may be of commercial value.

In 1976 John Moore's greatly enlarged spleen was removed as part of early treatment. Research revealed a rare variant of hairy cell leukaemia. A productive cell line was established, known as Mo-T. A complicated process of litigation ensued. Finally the California Supreme Court handed down a judgment which distinguished between Moore's cells and the cell line. Erin, however, does not pursue the question of whether Moore owned the excised cell tissue. Instead he sets out to argue that *a cell line is neither a part nor a product of the human body*. Thus, whether one starts with the premise that Moore owns his excised spleen tissue *or* with the premise that he does not is irrelevant. Either way Erin's argument will arrive at the same conclusion.

The argument runs that there were many vital ingredients in the mix which constituted the cell line. These include: Moore's tissue; culture media and laboratory equipment; high-level technical knowledge; procurement of safety; availability of esoteric knowledge; intellectual inventiveness and inquisitiveness over and above technical proficiency. Proceeding on the first premise noted above, Historical Entitlement Theory would argue that the begetters of the cell line must own *all* the items which go into the creation of Mo. Assuming Moore owned the excised spleen, the researchers could appropriate it by Moore's sale, abandonment or donation. But if we proceed on the second premise noted above, namely that Moore did not own the excised spleen, *the same conclusion applies*. The researchers were the first to acquire ownership of something that previously was not owned (the cells); the other requirements were supplied by the researchers themselves. Could it be argued, Erin asks, that society should deny to the researchers ownership of the cell line and establish it as a form of public property? That would depend on the judgment being made that the cell line was most beneficial to society when fully located in the context of biomedical research. However, to safeguard society's interests in the cell line, it may be necessary to impose restrictions upon the rights of the owners.

If the law were to change so that individuals owned their excised tissue, then a legitimate market would be required in which it would be proper that patients and research subjects should share in profits made. Certainly, many problems would flow from such a development. On the other hand, where a clear social benefit can be demonstrated, may not the cell line be derived regardless of the individual's property right?

Erin concludes that, in this actual case, it might be claimed that Moore has been morally wronged but not legally harmed by his lack of profit from his excised spleen. It is a matter for discussion whether there is a case for extending present legal property rights to take account of new scientific developments and ethical considerations.

In 'What "bugs" genetic engineers about bioethics' Peter Wheale and Ruth McNally use Anthony Giddens's characterization of late modernity, as presented in his study *The Consequences of Modernity*, as a heuristic device with which to examine genetic engineering. They argue that recombinant DNA technology has an important function in the transformation process of late-modern institutions (capitalism, surveillance, industrialism, military power) into those of a post-modern order.

The authors begin by supplying a short account of the history of genetic engineering to illustrate the discontinuous nature of recombinant DNA technology. In particular, it is the *micro*genetic (as they call it) character of modern genetic engineering which distinguishes it from its predecessors. Wheale and McNally then present Giddens's picture of the institutional dimensions of modernity and the way in which these can be used to analyse modern genetic engineering. They go on to explore the source of dynamism (time–space distanciation, disembedding mechanisms, reflexivity) lying behind these dimensions. 'Trust' and 'risk' are examined as characteristics of modernity, with their important implications for the social relations of genetic engineering.

The authors single out as another feature of modernity the existence of social movements whose formation is a response to an awareness of the high-consequence risks which modern institutions engender. These movements articulate what they consider to be the risks of the new technology and prescribe guidelines for alternative future transformations.

In conclusion, Wheale and McNally suggest, with regard to the modern bio-industrial complex, that 'the critics' demands and the proponents' promises converge on a future orientated along the contours of a realizable utopian post-modern order'. Signs of encouraging trends are visible in regulations which prohibit certain forms of genetic recombination, in the formation of new expert systems which are trusted to evaluate risk, and in the 'greening of biotechnology'. The authors are of the opinion that this kind of reflexive process shows a tendency to adapt current institutional norms to more humanistic and ecologically based values under the constraint of a future-orientated ethics.

Matti Häyry, in 'Categorical objections to genetic engineering: a critique', uses the conventional ethical distinction between consequentialist (pragmatic) and deontological (categorical) approaches. But he deals almost entirely with the latter, to see whether the moral status of genetic engineering can be established independently of the expected consequences of that activity. Häyry selects for special consideration somatic cell therapy, germ-line gene therapy, and the human genome project. Against the consequentialist claim that these genetic practices are immoral because they are, say, physically dangerous, the deontologist would assert that gene technology is inherently immoral because it violates rules drawn up by, say, the human community or by God or by nature. What, then, might it mean to assert that interference with the human germ line would be a case of 'playing God'?

Proceeding from Ruth Chadwick's analysis of 'playing God', Häyry sees problems in Chadwick's preferred interpretation, namely that in matters of life and death we may justifiably feel that no one else is qualified to judge whether our lives are worth living. This interpretation is not by itself sufficiently strong to refute any actual practices. If there are fixed moral limits, who draws them and where? Häyry rejects theistic accounts, accounts based on human destructive power over the natural environment, and accounts relating to a general sense of threat posed by technology-related actions – as these are all set out in Chadwick's helpful categorization.

What, in this case, of the formulation of genetic engineering as 'unnatural' or 'against nature'? The report of the German Enquete Commission stated that the development of individual human beings is natural only *if it is not determined by technical production or social recognition*. In practice somatic cell therapies performed on foetuses, infants and adult human beings are, in Enquete's terms, justifiable because the individuals in question have already developed into the beings they 'naturally' are. On the same principle, the mapping of the genome is licit for diagnosing the need for somatic cell therapies, but illicit for eugenic purposes. Finally, cloning, large-scale eugenic programmes and germ-line gene therapies should be prohibited, since they diminish the independence and uniqueness of human beings.

Häyry counters, first, that if we condemn genetic engineering because of its power to change, then most medical interventions should be condemned as well. Second, few philosophers would defend Enquete's strict biological definition of personality and individuality. Third, the proposition that genetic engineering would undermine the worth and dignity of human beings thus altered or cloned may be insulting to them. But this proposition is in fact not deontological but consequentialist, and so does not *require* the rejection of these genetic practices. In any case, Häyry argues, people do not need to know who is genetically altered, and there is no need to assume that people's attitudes to each other would depend on knowledge about each other's genomes.

Häyry concludes that the deontological arguments against gene manipulation largely lack merit. Those which have some value turn out to be consequentialist arguments. So 'The ultimate justification or rejection of biotechnology must be based on pragmatic considerations.'

In 'Biotechnology, friend or foe? Ethics and controls', John Harris examines two recent and influential attempts in the United Kingdom to set limits to the legitimate use of genetic engineering. In pointing up the deficiencies of these two attempts, he draws some general conclusions about the regulation of biotechnology and in particular about the effectiveness of committees of 'the great and the good' in formulating public policy.

His essay critically examines the United Kingdom Government's report – *The Report of the Committee on the Ethics of Gene Therapy* – and that of the British Medical Association. He suggests that in similar ways, both these reports are vitiated by unexamined prejudice and have reached conclusions unwarranted by the weight of argument and evidence assembled within them. In particular Harris rejects the supposedly 'ethical' distinction both draw between germ-line and somatic-line therapy.

Harris believes that the benefits of a 'slow but sure' approach to scientific advance have to be balanced against its costs. These costs, when they are measured in terms of lives ravaged by genetic disease which might have been prevented or cured, require us to justify doing nothing with the same care as we justify doing something.

In 'Genetic engineering and ethics in Germany' Ursula Wessels does not ask how one *should* go about answering the ethical questions raised by genetic engineering so much as describe how these ethical questions *are in fact* being treated in Germany today. The author first outlines the primary legislation, viz., the 1978 *Regulations*, with subsequent modifications, and the 1990 Law to Regulate Questions of Genetic Technologies. Some find the 1990 Law too permissive, and some find it too restrictive. Wessels begins with the debate about somatic-cell gene therapy. This procedure gains wide assent. Two contrary arguments are, first, objections to reductionist medicine, and second, the need for giving priority to preventative measures. The large majority judges that the procedure is *a* form of substitution therapy and thus ethically unproblematic.

Contrariwise, there is a large majority against germ-line gene therapy. There are several types of objection. First, embryo experimentation would have to proceed in order to establish the therapy; but embryo experimentation cannot be justified and is now in fact illegal. Second, there is a very general appeal to 'human dignity', to not treating potential persons as 'mere' means to ends. Third, a slippery-slope argument applies here. Fourth, the argument from unpredictable consequences is used to some extent. The author next deals with arguments about eugenic engineering. This is rejected even more vehemently than germ-line therapy. Wessels reports evidence for a good deal of rhetoric and little reasoned argument. Arguments, where these are found, are similar to those against germ-line gene therapy. There is a general appeal to the 'inviolability of human nature'.

Wessels concludes that: proponents of these procedures are in a small minority; their arguments are very like those of English-language philosophers; Germany is more united than most against genetic engineering; and less than precise appeals to the Nazi past are common.

Over recent years, not only secular philosophers, but also theologians and theological ethicists have been active in the debate about biotechnology. In 'Genetic engineering in theology and theological ethics', Anthony Dyson submits a number of pertinent texts to critical analysis. The texts comprise two putative 'extreme' pieces by Paul Ramsey ('Shall we "reproduce"?') and Joseph Fletcher

('Ethical aspects of genetic controls: designed genetic changes in man'). Within those outer limits, three World Council of Churches reports are explored (*Ethical Issues in the Biological Manipulation of Life*), (*Experiments with Man*), (*Manipulating Life: Ethical Issues in Genetic Engineering*). The WCC has maintained a steady flow of well-informed material on the biotechnological issues. In addition, a Church of England report (*Personal Origins*), stimulated by the Warnock Committee, is investigated.

A number of informal criteria are briefly applied to the technological texts. These include: past societal record; political and social self-consciousness; theory and praxis; ethics of experimentation; present and future generations; natural-law theory; boundaries between ethical and unethical therapy; nature and artificiality; theological revisionism; 'playing God'; somatic-cell and germ-line therapy.

Dyson is eager to characterize the overall theological tendencies in the texts. In particular he looks at these tendencies with a view to ascertaining how 'theology' relates to other interested disciplines. Can theology take part in a public debate about biotechnology, or is it confined to serve as a language for 'committed' religious groups?

1

MODERN ERRORS, ANCIENT VIRTUES

Stephen R. L. Clark

SETTING THE SCENE

Biotechnology is the art of manipulating living forms as though they were machines. We have been manipulating, and transforming, living forms since we adopted pastoralist ways – by breeding, domestication, training – but it is only recently that anyone has supposed that we could alter outward forms or behaviour by interfering with the inner mechanisms, the mechanical, biochemical and genetic processes that sustain outward shapes and motions. In the past we could do little more than select parents with desirable characteristics in the hope that they would engender what we wanted, and – once the offspring were there – punish or cajole them to do what we wanted. Biotechnology offers the hope that we could achieve a more secure result by altering the biochemical base on which the characters we desire are founded. It is not clear whether, once the possibility is realized, we could retain our present conceptions of human life. If we could engineer saints, heroes, savants, slaves or psychopaths to order, would any of us be *human* any more?[2]

But that issue is not my present concern, except indirectly. If biotechnologists are ever to succeed, they must engage in invasive experimentation on living systems. Eventually, they will experiment on 'human beings'. But their chief victims will at first be non-human, even though the drift of their endeavours must be to transform human beings into the very things that we have reckoned merely animal, of whom Aquinas said that 'they do not act, but are acted on'. Before we allow that denouement we should pause to reconsider what our right relations with the animals on whom we – and biotechnologists – rely might be.

The topic of animal experimentation arouses many passions. On the one hand are those who see it, straightforwardly, as torture, the deliberate inflicting of pain on defenceless creatures. On the other, those who as passionately believe that we must go on learning how animal organisms work if we are to continue to discover cures for disease and disability. Some of the latter, historically, have held that 'animals' hardly feel at all, or (if they do) it doesn't matter much. Others hold only that some pains are unavoidable if science is to progress, but that the scientists concerned will certainly be doing their best to ensure that as few and as trivial pains as possible are caused to our unwilling collaborators. The most recent British Act on the treatment of experimental subjects in the United Kingdom requires that animals judged likely to be suffering intense and long-lasting pain be killed. Often enough, neither side can quite believe that the other can be serious. Fox-hunters typically believe that hunt saboteurs are communists or worse; saboteurs typically believe that hunters are bloodthirsty yuppies or worse. So also here: experimenters think protesters are sentimental Luddites; anti-vivisectionists believe that scientists are career-conscious sadists. Both sides can *occasionally* uncover reason to suspect that they are right.

But the issue can be approached academically, with hatred and contempt for none, and it better had be if we are ever to be able to approach it politically as well. Such academic treatment has its own drawbacks. It must take place within a shared tradition of reasoned argument, and that very tradition may, historically, have built-in limitations. Patriarchalists can often sound to themselves like reasonable people just because it has been axiomatic that reason is a masculine preserve. Feminists who set themselves to argue with the enemy sometimes feel themselves at a disadvantage if they have to do so in a masculinist tongue, and respond by denouncing 'reason' as a patriarchal tool. Much the same problem faces radical zoophiles: it is 'obvious' to any reasonable person that 'animals' matter infinitely less than 'humans', and anyone who doubts that this is true cannot be serious. I remain convinced that Reason transcends tradition, even though it must take shape therein, and that we ought always to suspect what 'reasonable people' say.

Experimentation is not the only – or even the most significant – area in which our relations with non-human animals need to be reconstructed. I have already mentioned parallel disputes about the rights and wrongs of hunting. Hunters in these settled islands insist upon the need to control animal populations and the relative humanity and success of hunting with hounds (as opposed to trapping, poisoning or shooting); they also associate themselves with hunters in quite different milieux, for whom hunting is a necessary stay against starvation. Those opposed to such local practices doubt the sincerity of spokesmen for field sports in making those associations: shooting and trapping somehow become humane when they in their turn must be defended against complaint; the needs of hunter-gatherers are not those of the Home Counties. But hunting, shooting, fishing (however significant they may be as totems) are also not the major cause of animal exploitation or employment. It is not unreasonable for huntsmen and experimentalists alike to comment that their victims are numbered in hundreds or thousands, while the victims (direct and indirect) of our agricultural and domestic cruelty are numbered in many millions. Huntsmen and experimentalists constitute easily defined targets: those who finance what radical zoophiles consider cruelty in and around farms are still most of the population. Even our casual kindness and projective sentimentalism, exercised on furry creatures close at hand, is often exploitative – a fact we find as difficult to face as patriarchalists do the oppressive quality of their sentimental 'respect' for women.

The point of these reminders is just this: experimentalists can usually insist that they are as likely to obey the unspoken laws of humane life as anyone else. There is little reason to suppose that there are more cruel and negligent people in the scientific world than anywhere else. There may be fewer sentimentalists, simply because people unable to control immediate impulses of unreflective kindness or distaste will not go far in the profession. But the standards scientists work to will be very much like those of society in general. Why, then, should they be condemned? A partial reply might draw upon the undoubted internationalism of science: the scientific community, properly enough, does not draw its standards only from one nation's mind-set. Britons in the USA, for example, are regularly reminded that their own (maybe sentimental) concern for animals is not widely shared. American children, quite apart from the pervasive influence of a sentimentalized brutality enshrined in Wild West stories, are educated in school biology to discount the thought that animals can suffer pain or boredom or annoyance. Such ill-argued behaviourism is much less of a cultural force in Britain, and our national myths are of a more domestic kind. There are, after all, forces that do incline the scientifically trained to discount more of their untutored or their cultured sentiment than others. On which more below.

But the chief point I wish to insist on here is this: it may well be true that scientists mostly act upon the very same moral assumptions as non-scientists, and that their treatment of animals is no worse (and may sometimes be much better) than others'. The radical charge against experimentalists is not that they are unusually cruel or negligent, but that they act out – in a particularly noticeable form – assumptions deeply ingrained in contemporary culture which are actually false or wicked. Those who are seriously and deeply concerned about the whole practice of animal experimentation (and not just about occasional and obvious abuses) challenge human culture.

II THE IMPORTANCE OF SCEPTICAL ENQUIRY

So what are these assumptions, and if they are false (or even obviously false) how can they still be influencing us? It is important to remember that we are all inclined to go on acting on ideas that we actually *know* are false. One of the most difficult of lessons to learn is simply to eject hypotheses that we have found are false. You may find that surprising: surely we are all honest and reasonable people here, quite ready to try out hypotheses and abandon those that we have found are wrong? Of course there will be some hypotheses of which we are so fond that we will find it hard to agree that they *are* false. But once the point is made, that so they are, of course we shall abandon them. To say and believe that they are false just is to abandon them.

Unfortunately things are not quite so simple. An idea once entertained has its own influence. We are all bedevilled by hypotheses long since proved false, that we do not wish true, which yet control our actions and beliefs. The trouble often lies with the very principle of non-contradiction, that is the principle of rational thought. We do not want to contradict ourselves, and for that very reason cannot quite give up any idea, however foolish, that we have once given room to. We prefer to devise all manner of epicycles and *ad hoc* inventions rather than admit that we were, simply, wrong. But even that is not the whole explanation. It just seems to be the case that we find any sort of spring-cleaning very difficult. Psychologists have tried the following experiment: the subject population are asked for their opinion on some leading figure of the day, and then told a story very much to that person's discredit. Obviously enough their opinion of the figure goes down. They are then told that the story was entirely, absolutely, false, and their opinion sought once more. It remains far more unfavourable than it was when they began. The moral, simple-mindedly, is that mud sticks. More generally: we are often moved in ways that rationally we could not defend by the lingering effects of stories that we know or believe are false. Without an occasional dose of deliberate, sceptical enquiry, we shall continue to misdirect ourselves.

Such scepticism is a philosophical tool, and philosophers, like everyone else, are all too inclined to relapse upon the certainties born of beef and backgammon once the mood has passed. Perhaps that is inevitable, and I must accept that most of my readers will regard my attack on our present presuppositions as deliberately extreme, a device to be accepted as a way of pulling us back towards the 'norm' of *moderate* concern for animals. I doubt myself if there is any real content or foundation for that norm: my scepticism lies deeper, however uncomfortable even I may find it. Consider it possible, please, that we have all been wrong, and that the way to truth requires a really radical repentance. Unless you take that seriously you will not even gain the advantage of a moderate dose of scepticism, and still be left enacting ideas that you know are false.

III EGOISM

The first and obviously false assumption is egocentrism. It is obvious (who could seriously doubt it?) that there are entities distinct from me, who carry on existing when I do not think of them, who owe me nothing and do not think of me at all. A true account of things will not pick Me out as something special, will not set the worlds in orbit round my centre. The true and only centre of the worlds is everywhere, and its circumference nowhere.[3] Yet every day I act as if the things I do not know about (or choose to neglect), from the dripping tap to the dying Ethiopian, do not exist. Only the most rigorous employment of a sane imagination allows me to admit the obvious: that those who die in pain are no less pained because I do not think of them, that my friends exist when they are out of sight, that what I do not know about at all may yet be real.

Practical solipsism is closer to us than we usually suppose, and there have even been philosophers, and very distinguished ones, who have endorsed a full-blooded solipsism. But most of us are probably too nervous to stand quite alone! The nearest we can get to sheer, unashamed egoism is to identify ourselves with clans or castes or companies against Them. Moral behaviour begins in loyalty to just such groups, just such an insistence on 'doing good to our friends and harm to our enemies'. Our 'friends' are simply those with whom we belong, and who belong to us. If a supposed ally, one of Us, is caught considering the interests of Them (as it might be injured Argentinians or class enemies or animals), We are yet more enraged. Those who give Them any thought at all are judged to have betrayed Us, and to be secretly enamoured of the alien ways We impute to Them. We know all this for nonsense, but still act as if it were true. *They* are 'animals', in the sense beloved of judges: uncultured egoists eager to satisfy their every impulse. Real animals are seen as symbols of the alien ways we impute to Them in considering Them animals. So anyone who wishes to give real animals a hearing must be on Their side.

But try to consider things as they would appear to an ideal observer, or to the Creator-God: why should, say, a dog, a rat, a monkey be required to live a tedious and imprisoned life, and die an agonizing death so that a hominid be spared the least discomfort?[4] Because we are hominids we may prefer (egocentrically) that non-hominids pay the price of living, but how could we persuade any ideal observer, God or disinterested jury that our preferences have more weight than those of the dog? Why isn't that just like saying that my personal comfort must be more important, absolutely, than your vital needs?

IV HUMANISM

The second principle behind our doings is humanism, which purports to offer some reason for that prejudice that would or should persuade an ideal jury. The humanist revolution demanded that we attend to all members of our species, whatever their particular relation to us may be. Whereas pre-humanist societies endorsed a rigorous division between Us and Them, between our fellowship and foreigners, humanists believe it obvious that all (and only) human beings deserve the same respect. Foreigners are of no other kind than I, and all our conspecifics differ from non-human beings in deserving an absolute respect. Old-fashioned liberals identify the 'rights of man' as rights to life, liberty and property. It would be wrong on these terms to kill, control or dispossess any human being. Liberals in the sense now generally meant impose more positive duties on us all, of care for the afflicted: human rights include the right to be fed, clothed, educated and cared for (even to have annual paid holidays). In practice even such modern liberals do not admit to any enforceable duty so to care for more than their fellow-nationals; and old-fashioned liberals, regrettably, have not always recognized the claims to liberty and property of those they reckon 'savages'. But humanism, in libertarian or welfare forms, has played a part in the extension of existing rights and privileges from members only of our class or sex or nation to the whole species. Where unembarrassed groupies (so to speak) think it only right to pursue Our interests at whatever cost to Them, humanists insist on taking account of Theirs, and reckon it absurd to think that what is bad when done to one of Us is somehow quite all right when done to Them.

Unfortunately for those outside the magic circle, humanism disallows even such paltry rights as had been conceded to non-humans. Hindu respect for 'sacred cows', or Jain disinclination to kill or hurt any living creature, or 'primitive' respect for 'totem animals' are easily equated with sentimental concern for pets: proofs that the animalist is personally inadequate, starved of human affection and neglectful of their 'real' human duties. Such humanism, of course, may also insist that certain styles of human living are unworthy of the name: sadistic enjoyment even of non-human pain, or negligent treatment of the animals one owns, are ways of not living up to the humanist's ideal. But it has never been easy to explain quite why such active cruelty or negligence is wrong, if only human interests matter to the rational mind. If humans matter so much more than non-humans it must be that they are of so radically different a kind that it is hard to see that 'hurting animals' is of the same kind as 'hurting humans'. Perhaps the sadist's error just is to suppose that he/she is *hurting* animals – but negligence or indifference are then not crimes.

This is the kind of humanism often deployed by experimentalists. Ethical considerations preclude invasive experimentation upon 'human beings', but not on 'animals', because all and only human beings merit rational respect: the most that animals can expect is sentimental interest of a sort that should not stand in the way of human interests. Animals either don't feel pain, or boredom, or distress, or else the 'pains' they feel are of another, and less interesting, kind than 'ours', because 'we humans' are another sort of thing, objectively, than them. A report some years ago from China suggested that biologists there were seeking to produce a human-chimpanzee hybrid which would be available as an experimental subject when 'ethical considerations' forbade the use of truly human stock.[5] Chinese authorities, of course, show little enough concern for *human* rights, but the same strange disjunction features in western ideology as well. It is only pure *human* nature that requires respect. Why?

One version of humanism is no more than the near-solipsism that I described before, and so no real answer to the question why God should prefer to hurt non-hominids. Imaginary cases may help to distinguish this variety from the more 'objective' kind: merely 'subjective' humanists do not suppose that intelligent extraterrestrials need, in moral reason, to give any weight to human interests over those of terrestrial organisms more to their taste. The medieval knight who preferred to save a lion rather than a snake, 'because it seemed the more natural creature', perhaps expressed a similar subjective preference for fellow-mammals. Betelgeusian arachnoids might well prefer spiders, and 'we', correspondingly, need feel no qualms about the arachnoids, however civilized they were. On these terms, humanists only urge us to identify with human beings and things that look like human beings: creatures that have faces benefit from stray affections. 'Objective' humanists think otherwise, that all and only 'rational' beings (or beings of the same real kind as rational beings) deserve respect. The arachnoids ought to be treated well, and should, in reason, treat 'us' well. What matters is the mind behind the face.

Both sorts of humanist usually make two assumptions: first, that a species is a natural kind (a notion that now has very little biological backing), and second, that the human species is the unique embodiment (on Earth at any rate) of sacred Reason (a notion that almost no philosophers – other, I must admit, than theists like myself – now take seriously even if they still endorse humanistic values). After all, not all our conspecifics are actually rational, nor are they all so easily identified with. Some moralists are ready to write off imbeciles, or human beings too unlike themselves, but humanism, in whatever form, still demands that *all* our conspecifics have equal rights (whatever they are). So subjective and objective humanists alike must include in their concern all such as are of one kind with the rationally active or the sympathetic few. It is that very notion of 'one natural kind' that is now suspect.

A biological species is not now thought to be a set of organisms with a shared, essential nature, such that there is or could be an organism 'typical' of the species and that other organisms may be more or less 'defective' specimens. That notion of species-membership is one that modern biologists usually call 'Aristotelian', although Aristotle himself was almost certainly not guilty of the crime.[6] The modern notion is rather that a species is a set of interbreeding populations, and that no single specimen is more or less what the kind should be than any other. What is called the type-specimen, in textbooks, may turn out to be very 'untypical'. There is no need to think that all and only species-members have any shared character, save only that they have common ancestors and are still members of a population whose members may yet have common descendants. The matter is further complicated because a good many terms of our folk-taxonomy (like fishes, creepy-crawlies, weeds) turn out not to name scientific taxa. There are, in a real sense, no such things as fishes (for what we call 'fishes' may belong to radically unrelated groups). It is not long since scientists doubted that there were even humans.

The fact that we share ancestors and may share descendants perhaps gives some weight (psychologically) to subjective humanism. But that sort of humanism has no stronger claim than loyalty to any natural group: because our common ancestors are further off, and our shared descendants may never in fact exist, subjective humanism has less weight, in general, than closer ties. If it's all right to be partial to the claims of humans over non-humans, it must also be all right to be partial to the claims of human beings who are more like us, or better known to us. But that is to abandon the high moral ground on which humanistic moralists prefer to stand. So those who think that humanism has more moral weight must probably endorse the objective form: but that is to neglect the biological discovery. Even if all actual humans really did share some important character deserving absolute respect (say, Reason), nothing says that all and only humans must, any more than it is bound to be true that all and only reptiles have three-chambered hearts. And there is evidence, in any case, that not all our conspecifics are thus rational, and that others may be.

Consider the following analogy. 'Mammalism', so to call it, does play a part in the choices that we make: furry creatures with faces get a better deal than scaly ones (unless they are rats). But even if we elevated this subjective preference to the point of objective principle (as it might be: all moral beings must respect all creatures that are cuddly and care for their young), it would surely be odd to deny respect to non-mammals that did the same, and respect what are technically mammals even though they didn't. It would be odder still to go on being mammalists when we had found out that, strictly, there were no mammals (any more than fishes).

In brief, subjective humanism makes no general claims about the sort of things we should respect but (for that very reason) lacks the moral force regularly imputed to it. If humanism is anything more than the weak claim to mind about all those with whom we might share descendants (but not as much, no doubt, as we do about those with whom we do, or realistically might), it must claim that there is an objective duty so to mind. But what is there to mind about that all and only human beings share? Not all are really rational, and some non-humans are (by any ordinary standard) quite as rational as humans mostly are. If we have a reason beyond group-preference for wishing humans well, then it will apply to non-humans too. And it remains very unclear why an ideal observer, God or disinterested jury would find the capacity for rational thought (whatever it is) so overwhelmingly important. Is it simply that any such jury must, if they are to judge at all, share those rational capacities, and so be compelled to respect them – and no other – on pain of denying the value of their own power to judge? It does not seem to me to be a strong argument. After all, if they are to judge at all they must also share – and respect – capacities, such as sentience, appetite and affection, that non-human beings certainly possess.

V UTILITARIANISM

The third principle underlying much public moralism is utilitarianism. The term has many meanings by this time, including ones the movement's founders would despise. Utilitarians, to state what should be obvious, do *not* judge actions, people, architecture by their mere 'utility', the efficiency with which they serve some mercenary end. Real utility, in the sense utilitarians meant, includes the increase of pleasure (of all kinds, but with a learned preference for the more sociable and intellectual pleasures) and decrease of pain. Beauty is a utilitarian goal.

The nineteenth-century movement helped to secure some gains for animal welfare. What mattered, as Jeremy Bentham said, was not whether animals could talk or reason, but whether they could suffer,[7] and the idea that they did not – however popular with animal experimenters from Malebranche to Claude Bernard to . . . any number of pontificating moderns – has never satisfied anyone without something to gain. Non-human animals have nervous systems and produce much the same natural opiates to counter pain as human animals. It is easy to believe that they mind less about their pains than we: just as it is easy to believe that the poor mind less about their troubles, or that Third World peasants do not need the medical care that sensitive souls like us can't live without. But historical ignorance and egoism are easily recognized too.

If pain is an intrinsic evil, something that anyone would wish to eliminate, then animal pains are too. Some versions of utilitarianism seem to speak of pain (and pleasure) as a sort of stuff, to be eliminated (or increased) no matter where or how the job is done. This crude and unreasonable theory would excuse our causing very great and lasting pain if thereby the notional total of pain were reduced or the total amount of pleasure (housed perhaps in an elite) were increased. Unfortunately for the animals in our charge, it is just this sort of utilitarianism that rules in too many experimenters' hearts. A movement that began by demanding moral respect for animals (and that was severely criticized by orthodox humanists for doing so) has since been used to defend the use of animals in painful and oppressive experiments: only so will the sum total of suffering be reduced, or the pleasures of the elite secured.

Utilitarian calculation of this kind has one stunning disadvantage (quite apart from any moral obloquy it may incur): it is wholly incalculable. Who can tell what the sum total of suffering may be or what might decrease it? Who can tell what effect a rising (human) population has? Is it a mistake to save a life if it turns out that the survivor suffers later on? Or suffers how much? Or costs the whole how much? Sensible utilitarians do two things. First, they cease to pretend to calculate the exact effect of any individual act ('act-utilitarianism') and instead act on such rules as acting on produces best results ('rule-utilitarianism'). Second, they abandon talk of pain or pleasure as quantifiable stuff, and instead begin to talk of the real or rational preferences of the creatures affected by our acts. 'Each to count for one and none for more than one': everyone affected by the rules has as much say in what they should be. What rules, then, could or should 'we' agree on? What rules will 'we' be 'happiest' with?

I said that sensible utilitarians do this, but the truth is rather that by being sensible they cease to be utilitarian. For what we seriously prefer may not, in any clear sense, be 'what gives us most pleasure and least pain' (unless we define 'pleasure' as getting what we want). And we have no conception, prior to the votes we cast, of what would make us happy. It's not as if we knew what happiness was, and only wondered whether we'd be 'happier' as pirates or as priests, or whether a world of pirates or contented wombats would be 'happier' than one of peasants or hunter-gatherers or biotechnologists. All forms of life have their own drawbacks and delights. The utilitarian calculation turns out only to be an extravagantly General Election, and the result will turn on the prior moral choices of those allowed a vote. The first kind of utilitarianism, however absurdly, identified a goal (that world is best that has most pleasurable sensation for least cost of pain) which differed from other real and imaginable goals (as it might be, one of equable beauty or of romantic anguish or of duty well performed). The second only offers a way of solving practical disagreements: we get the world most voters want enough. Gang-rape is only outlawed if more voters are more outraged by it than are delighted or indifferent. If enough voters already think it wrong to hurt animals it may be right to prevent such hurts; if they don't, it won't. But this tells us nothing at all about which way to vote.

Sensible utilitarians have usually lost sight, as well, of what both Bentham and Mill insisted on: that animals are voters too, or must be supposed to be. If the best outcome is the one that satisfies most voters best, and no one counts for more than one, it must seem very doubtful that the present state of things is best. How would our animals vote if they were allowed the chance? What they would vote on, remember, must be the rules under which we live. Would our victims seriously prefer to live by rules that gave their exploiters the unquestioned right to imprison, kill and torture? What advantage could they derive? One answer, that they thereby gain or keep existence (since they would not be born in such numbers if we did not intend to imprison, kill and torture them), presumably rests once more upon an unexamined use of 'totalizing utilitarianism'. What advantage does the existence, under deeply unpleasant circumstances, of large numbers of her kind give to any actually existent creature? Would any rational being choose to produce descendants as numerous as the stars of heaven if the price were that she produced them by mechanized rape, and lost them to the slaughter house or to be similarly raped, even before a natural weaning? Would it make any difference even if these descendants had occasional pleasures of a kind and intensity that free animals could not?

One further feature of utilitarian theory: whereas our ordinary maxims still distinguish acts of omission and commission, a mere concern for consequences sees no difference. If an act of mine results in someone's pain or death it makes no difference whether I directly caused that evil, or merely failed to prevent it. So experimenters claim that failing to perform admittedly invasive experiments will, over the long run, cause more evil. The argument is a lot less certain than they think, simply because we cannot predict what scientific progress *would* be made if we decided not to use such methods, and because few scientists have the nerve to admit that experiments on *human* beings (against, of course, their will) would – on their terms – be likelier to result in progress. We renounce such imagined profits in the latter case, and so cannot, without additional argument, insist upon them in the other. But the principle on which they are relying is itself a suspect one, not least because it has profoundly totalitarian associations. Negative duties, not directly to cause evil, are more universal and more powerful than the positive ones, to prevent evil or cause good. At the least, there are sound reasons for us to make distinctions – but the argument here would take too long to unfold. It is at least counter-intuitive to suggest that by failing to write a cheque for five pounds which would then be used to save an Asian peasant's life I am as guilty of her death as someone who guns down that peasant. Indeed, since the latter killer presumably gets more pleasure out of it than I do from my negligence, I am, on crudely utilitarian grounds, guilty of a greater wrong. This seems implausible, but is the image of experimentalist utilitarianism.

Some recent philosophers have attempted to argue from utilitarian principle to the conclusion that we are indeed doing wrong, and should cease torturing and killing animals. They write in Mill's and Bentham's spirit. But the very fact that other philosophers have defended the status quo on just the same principles (for any alteration in the way we live will have unpleasant consequences for those currently employed in farms, laboratories and zoos) reveals that the calculation is, as I said, incalculable.[8] Professed utilitarians only *think* they reason to their conclusions by the calculus: in fact they choose their goals by other means. Utilitarianism is always and entirely an excuse – and one most used by the exploiters of our non-human kin. It may seem obvious (so many things seem obvious) that utilitarians should – for example – engineer food-animals that positively enjoy being eaten, and can tell us so.[9] But in what sense are the inhabitants of such a world 'happier' than those who refuse to see their fellows just as food? Is even the 'Dish of the Day' *happy*?

VI OBJECTIVISM

The fourth principle that affects our present reasonings is Objectivism. Groupies take it for granted that what matters about things is what We call and think of them. Weeds are there to be uprooted; pets and pests and pigs are different because We give them different values. It is a profound and valuable discovery that we can look past such subjective judgments to consider what things are in themselves, what they would be for an omniscient, free spirit. The virtue of objectivity just is the willingness to put our prejudice aside, and see things 'as they are'. Humanists (or objective humanists) put aside the merely conventional distinctions, of class or caste, in favour of more deeply rooted 'natural' distinctions. 'When Adam delved and Eve span, who was then the gentleman?' By this account we should regulate our feelings and behaviour not by conventional and changing labels, but by the real divisions of things. The trouble is, as I have pointed out, it is now very difficult to think that the human/non-human division is as clear or all-important as our predecessors thought. If we ought to treat relevantly equal cases equally we maybe ought not to do to chimpanzees what we refuse to do to human beings.

The problem is compounded by utilitarian infections. The original utilitarians thought it a matter of fact that pain was evil, and that we ought, having the power to do so, to alleviate or prevent it. But the drift of utilitarian argument is towards the thesis that 'there's nothing right or wrong but thinking makes it so'. What's wrong is what 'we' disapprove of. Emotivism is the meta-moral theory (the theory about the meaning of moral judgment) that fits the moral practice of utilitarians. But emotivism simply is the thesis that moral divisions are not matters of fact to be discovered by objective thought. An 'objective observer' does nothing, and does not care one way or another. In which case objective humanism and objective utilitarianism are both false. Preferring human beings, and preferring pain-free lives, are historically grounded prejudices.

Objectivism, which was once the firm intention to think real differences of more importance than conventional ones, has now become the intention to discount all moral values in reporting facts. In which case personal preference must rule in practical affairs, and we relapse, in practice, to the groupies.

The rhetoric of objectivism in its modern form has many disturbing aspects. Utilitarian calculation and objectivism together make it difficult to see why we should not use orphan or abandoned human neonates for our experimental purposes, breed pretty imbeciles for licensed pederasts and eat aborted foetuses in expensive restaurants (my apologies to all right-thinking readers for inducing nausea, but it is important to remember what the options are). It is doubly unfortunate, of course, that some zoophiles, intent on destroying humanistic values, have found themselves suggesting that such acts would be less wrong (might even, on some utilitarian calculation, be more right) than present treatment of non-human animals.

Objectivism in its modern form rests on very dubious arguments. My objection to it is not simply that there are real evils, just as my objection to utilitarianism is not simply that it neglects real evils. Utilitarianism in all its forms is quite absurd. Modern objectivism, or emotivism, is as well. I have found that experimental scientists are especially vulnerable to its influence, but are rarely acquainted with its history or sophistical roots. Scientists, they say, are only concerned with 'facts', and moral preferences can never be grounded upon facts, or rationally discussed. This does not seem to stop them thinking their own preferences more rational than those of zoophiles, while still denying that any such preferences are rational at all.

But what does ground the claim that there 'are no moral facts', and hence that moral judgment is only personal preference, and hence that scientists are somehow entitled to demand, as of right, the liberty to act upon their preferences? The arguments most commonly used are as follows. First, it always does make sense (of a sort) to deny that any particular ethical conclusion follows from any agreed factual premise. We can agree that an experiment causes pain but not agree that it should be avoided; we can agree that it may bring great advantages to the human species but not agree that it should therefore be performed. But if propositions of the form (p & q) and (p & not-q) both make sense, there can be no conceptual tie (of the kind obtaining between 'Joe's a bachelor' and 'Joe's unmarried') between p and q. So the connection between the 'facts' commonly held to support a particular moral judgment and that judgment is not a conceptually necessary one (if it were, the denial of it would make no more sense than the claim that one could be a married bachelor, or a round square). The only alternative generally allowed is that it be an empirical discovery that acts of such and such a kind are, as a matter of fact, always or practically always wicked. But 'being wicked' is a property quite unlike other and non-evaluative properties: it seems not amenable to laboratory controls, nor can it be recorded upon tape, nor is it transformable (in mathematical terms) into some other sort of property. We may one day discover how electromagnetism, gravity, the weak force and the strong are variations on a single theme, such that a single formula accommodates them all. But not even the most sanguine physicist believes that moral qualities could be incorporated in such a universal synthesis, as if it could even be true that faster, hotter, less acidic things alone were virtuous.

So moral laws (if they exist) are neither truths of logic, nor satisfactorily testable generalizations waiting to be subsumed in a unified general theory of the kind sought out by physicists. If those were the only alternatives, we should have to conclude (and scientists regularly do) that there are no moral laws outside the conventions of the tribe. By the same token, of course, the claim that only laws of logic and empirical generalizations have any rational status has no rational status: for it is neither a law of logic nor an empirical discovery. In denying that there are any real moral duties, scientists abandon any right to demand logical rigour, intellectual honesty or openness to refutation. If there are no moral duties then a preference for finding out scientific truths is no more than curiosity. Logical positivism (to give this theory its old name) is not well thought of by philosophers, but retains its influence in non-philosophical circles.

A second argument against the existence of *moral* facts gets its strength simply from the way we identify 'real facts'. A fact, in contemporary rhetoric, is something that just anyone must admit who approaches the data with the 'right' attitude. The right attitude in turn is understood to be one cut off from ordinary emotional responses: 'stick to the facts' means 'don't reveal your ethical opinion or your personal preference'. So facts are defined, from the beginning, as truths without an ethical dimension. Unsurprisingly, such facts carry no ethical implications. One might as well decide not to mention colours, and thence infer that all the world is colourless. But once again, the problem is that such a rule has no rational weight: how can we sensibly suppose it to be a matter of *fact* that we *should* take up this detached attitude, that we should say only what anyone could as easily say, rapist or Nazi or believing Christian? Why should those be the only facts? And how could they be? It plainly is not true that just everyone would agree they were: and so, if we should affirm only what just anyone would affirm, we should not affirm that. The objectivist identifies a *right* approach in the act of denying that there is a right approach.

A third argument against the truth of moral truisms is that people disagree even about truisms, let alone about detailed casuistry. Surely we would not disagree so much if there were a matter of fact to be discovered? But the truth is that there is much less disagreement about the truisms than objectivists pretend (though there is a lot about the application of the truisms), and that it is intolerably naive to think that matters of 'mere fact' are so easily settled.

In brief, the enormous influence of modern objectivism rests on very shaky foundations. Anyone inclined, like myself, to take an older objectivism seriously need feel few qualms: there is a right attitude to have if we are to see things clearly and to see them whole. That attitude does indeed require us often to put aside our idiosyncratic prejudice, to look past the creepy-crawly and find out the beetle. Moral reasoning itself often depends on trying to see things as they might be seen by anyone, and so to make no foolish exceptions for our 'friends'. But it does not follow that a claim cannot be factual unless it could be admitted to be true by anyone, no matter what their ethical intentions. Those who insist on seeing laboratory animals as mere 'animal preparations' or living test-tubes deny themselves the chance to see such truths as would cause difficulties for their treatment of them.

VII ANCIENT VIRTUES

I said that we were deeply affected by assumptions that we could easily discover were false or unreliable. Egoism, humanism, utilitarianism and modern objectivism are all principles bereft of rational support, and having most disagreeable implications. All four principles begin more plausibly. After all, the division between friends and enemies does run deep; the merely human form (irrespective of virtue, intellect or usefulness) is sacred in many traditions; pain is disagreeable, and pleasure worth pursuing; objectivity is indeed a virtue. But in every case a single useful principle is taken too seriously, as the only thought worth having.

I am conservative enough to think that we would gain a lot by carefully returning to the old tradition of ethical reasoning. Destructive scepticism is not the only worthwhile strategy. It is, to put it mildly, not quite sensible to refuse to believe things till they can be proved (either absolutely or to every sensible person's satisfaction). How could we ever begin or satisfactorily complete that project? We had better believe what we are told until we see good reason to reject it. Nor do I think it sensible to demand a single moral theory from which all right decisions can be easily deduced, as though we could ever relegate the pain of moral choice to pocket calculators. Morality is the record of 'our' past conclusions, the testimony of those who seriously sought to be virtuous, and to see things clearly. We ought, in general, to accept the moral truisms of our ancestors: not to do to others what we would not have them do to us, and to do for others what we would have them do for us.

So perhaps tradition is right, after all, to say that we owe less respect to animals than to humans, even if (as in many other cases) we can give no satisfactory account of that division or that duty? But tradition gives us no right to think that traditions are unalterable: refusing to change one's mind is exactly the intellectual sin I mentioned earlier. It is quite certain, for example, that our ethical traditions have given too little weight to women's interests; quite certain that the master–slave relationship has been endorsed at least since agricultural society began. There is nothing unusual, in that case, about suggesting that traditional views of animals are not merely ungrounded (which perhaps is no great fault), but demonstrably mistaken.

We are not of another, radically different kind. Ties of loyalty may bind us to particular non-humans as much as to particular humans. When we open our eyes to see the reality of another creature, and so learn to respect its being, that other creature may as easily be non-human. Those who would live virtuously, tradition tells us, must seek to allow each creature its own place, and to appreciate the beauty of the whole. It is because human beings can sometimes come to see that whole, and know their own place in it, that – in a sense – they are superior to other forms. Our 'superiority', in so far as that is real, rests not upon our self-claimed right always to have more than other creatures do (which is what our modern humanism amounts to), but on the possibility that we may (and the corresponding duty that we should) allow our fellow-creatures their part of the action.

Experimentalists, perhaps, do after all stand out from the normal mass of human error. I said at the beginning that there were other areas where more harm was done to animals, by far more people. But biotechnological experimentation represents our single greatest sin: the intention always to control what's going on, to evade all penalties. We experiment invasively on animal subjects because we hope to find out how things work and thereby evade the costs of pleasure-seeking and the pains of living in the real world alongside all our kindred. That impulse, to possess things perfectly, and never have to yield, is what was once identified as the first sin. It is one thing to mitigate the effects of obvious and immediate evils; quite another to attempt to 'remake the world entire'. The final irony is that in seeking to understand experimentally we break the thing itself. True wisdom, so tradition says, lies in acceptance.

Oddly enough, that very dictum can occasionally be found amongst the scientifically inclined: real scientists are often moved to enter on their profession, after all, by just that happy admiration of the real world and its inhabitants. It is a strange admiration that can so often turn to destroy its focus. Perhaps, when solipsism, humanism, utilitarianism and modern objectivism are recognized as absurd perversions of the ancient virtues, experimentalists may let themselves acknowledge what the best of them already know. Our duty is to admire and to sustain the world in beauty, and not to impose on others pains and penalties we could not bear ourselves. It follows, I am very much afraid, that we should not countenance the use of any creature in an experiment likely to impair its chance of living out its life in beauty. We ought to live by those laws that an ideal observer or Creator-God would make (maybe has made) for the world: to respect the integrity of every creature, and not to seize more for ourselves and our immediate kin than would be granted under such a dispensation.

NOTES

1 Originally written for a series of seminars on animal experimentation organized by The Centre for Social Ethics and Policy at Manchester University.

2 See my 'Notes on the underground' (*Inquiry* 33 (1990): 27–37), replying

to C. P. Ellerman, 'New notes from underground' (*Inquiry* 33 (1990): 3–26).

3 Which is an ancient definition of God: see J. L. Borges, 'Pascal's sphere', in *Other Inquisitions*, tr. R. L. C. Simms (Austin: University of Texas Press, 1964).

4 See E. Kohak, *The Embers and the Stars* (Chicago: University of Chicago Press, 1984), p. 92.

5 David Bonavia reporting in *The Times*, 11 December 1980. One good deed attributed to the Red Guards is that they destroyed the laboratory in which, it was alleged, Qi Yongxiang had aimed to produce a 'near-human ape' from a female chimpanzee inseminated with human sperm.

6 See my 'Is humanity a natural kind?', in T. Ingold (ed.), *What is an Animal?* (London: Unwin Hyman, 1988), pp. 17–34.

7 J. Bentham, *Principles of Morals and Legislation* (London, 1789), p. 310n.

8 See my 'Utility, rights and the domestic virtues', *Between the Species* 4 (1988): 235–46.

9 D. Adams, *The Restaurant at the End of the Universe* (London: Pan Books, 1980), pp. 92ff.

2

BIOTECHNOLOGY AND AGRICULTURE

David Colman

Technological change in agriculture has been the foundation of economic growth and development. By raising yields and continually increasing the number of people who can be fed by the output of one worker in agriculture, change in agricultural technology has permitted an increasing proportion of labour and other productive resources to be devoted to non-agricultural production. Plant and animal breeding and chemical and mechanical technology plus improved husbandry have caused continual structural change in farming. These technological stimuli have arisen as a series of overlapping waves to create a process of technological change, the momentum of which looks likely to be maintained by intensified adoption of information and biotechnologies. As a result of the cumulative processes to date, farms have become larger and more specialized, and farmers are more highly trained and have substituted machines for animal and human power; farming in the 'West' has become one of the most capital-intensive of industries. At the same time agriculture has become more dependent upon industry, and even more industrialized – the intensive production of eggs, broiler chickens, horticultural products and pigs has many characteristics which belong to industrial production rather than to traditional farming. Goodman *et al.* (1987) talk of 'the industrial appropriation of the rural processes'. Processes of marketing farm products have passed off the farm to be performed by sophisticated distribution and retailing sectors which increasingly dictate details of production to the farm sector, while at the other end of the chain an increasing proportion of inputs is supplied by the chemical, pharmaceutical and engineering industries.

The process of 'industrializing' agriculture has been transferred to the less-developed countries (LDCs) and is epitomized by the so-called 'Green Revolution'. This has entailed expansion of irrigation systems using modern pumps, engineered dams and canals, plus the increased use of inorganic fertilizer, insecticides, herbicides, fungicides and power machinery; although at its heart has been the 'old' biotechnology of breeding new cereal varieties which give high yields when supplied with chemical inputs. Because much of the impetus for this 'revolution' has been western technology and farming know-how, particularly that of the USA, there are those who are critical of what they see as the LDCs' increasing technological dependence upon the West, its industries and research institutions.

What is undeniably the case is that technological advance has been faster, and public and private expenditure on research in western agriculture greater than in LDCs taken as a whole. Thus exportable surpluses from the West have increased with generous public support, while many LDCs have had to resort increasingly to food imports to meet their needs.

While at this stage it is difficult to foresee exactly when, and on what commercial scale, the new agricultural biotechnology 'revolution' which is now brewing will have its effect, most commentators agree that it will intensify the industrialization of agriculture, and that it will increase the technological dependence of most LDCs upon western firms. It may cause considerable disruption to the economies of some countries and will cause farming operations to diverge increasingly from the pattern associated with the country yeoman. Rural life in many more remote farming areas will be further threatened by the continued deterioration in the economics of ruminant livestock farming on upland and low-productivity pastures. It is these redistributive and structural effects that are the central focus here rather than any attempt to estimate the scale and speed of uptake of specific biotechnology inputs and processes.

Although prospective biotechnological change in agriculture is not different in many fundamental respects from other processes of technological change, in that it will operate through the provision of new or modified inputs (seeds, hormones, etc.) and the creation of new markets in industry, it does raise new issues. Are there dangers from releasing life forms which are engineered using new methods which are quite different from those associated with varieties developed by traditional plant- and animal-breeding methods? Even if the dangers can be objectively assessed and turn out to be minimal, will what may be perceived as man-made and therefore 'unnatural' products be acceptable to society as a whole?

PRINCIPAL DIRECTIONS OF BIOTECHNOLOGICAL DEVELOPMENT

In a recent review of agricultural technology, the Office of Technology Assessment (OTA) of the US Congress (1986) defined biotechnology to include 'any technique that uses living organisms or processes to make or modify products, to improve plants or animals or to develop micro-organisms for specific uses – it focuses upon recombinant DNA and cell fusion technologies'. Longworth (1986) concurs with this definition, with the significant addition of tissue-culture techniques, an aspect of biotechnology which has great economic, and therefore social and political, potential impact. Despite debates about whether these or any other definitions are adequate (Buckwell and Moxey 1990), they do touch on the main aspects of biotechnology which deserve to be considered.

The aspect of this new biotechnology which most captures the imagination and stirs the greatest controversy is gene splicing or recombinant DNA techniques, which inspire researchers to consider the possibilities of producing reproducible animals and plants markedly different from current existing species, and referred to as transgenic species. Already in higher animals and plants recombinant DNA has produced transgenic forms which are being commercially exploited. Their agricultural significance is, however, so far limited, and commercial application is concentrated in highly profitable pharmaceutical and horticultural markets. Genes have been introduced into several animal species which alter their protein synthesis to enable transgenic sheep to produce insulin in their milk, and rabbits to produce interferon.[1] A completely different application has originated in Denmark for salmon, where it has proved possible to introduce germplasm which enables the salmon's physiology to handle heavy metals, which are normally toxic, so opening up new locations for farm fisheries. Much current research is directed to conferring disease immunity on animals, and holds out the prospect of widespread commercial application. In plants one achievement has been the transfer of genetic resistance to antibiotics in the petunia (Longworth 1986, p. 7), and another has been the introduction of storage-protein genes from French bean plants into tobacco plants. As yet commercial progress with recombinant-DNA technology in plants appears limited, but extensive opportunities beckon, particularly because plant research is less restricted by the ethical and animal-welfare concerns which apply to research on transgenic animals.

The most important commercial developments based on gene splicing have so far occurred with genetically much simpler micro-organisms, and it is with these that the greatest short- to medium-run commercial potential lies. Already genetically engineered micro-organisms are producing a variety of hormones, vaccines, enzymes and other proteins. Important examples are the production of insulin, vaccines for neo-natal diarrhoea in calves and piglets, and the bovine growth hormone BST identical to that produced in cows which can stimulate a 20 per cent increase in milk yield. BST (as discussed more fully below) is already a source of problems for legislators in the European Community, and has provoked strong reactions from the media and milk consumers.

As far as larger animals, and cattle in particular, are concerned, it is developments in embryo transfer and in many processes for manipulating reproduction which hold out the prospect of continuing increases in yields, feed conversion efficiency and general economic efficiency. Already, apparently over 1 per cent of dairy calves in the USA are from embryo transplants, despite the high costs still associated with this procedure. Widespread adoption of these sophisticated technologies would further distance livestock farming from its traditional rural simplicity, and from the natural mating of animals to produce offspring. It would place technical demands upon the operators which favour large-scale company farms capable of supporting a range of highly trained specialists.

Longworth (1986, p. 8) identifies new techniques of tissue and cell culture as having 'the potential for enormous advances in crop improvement in the next couple of decades'. Those techniques 'can both increase the genetic diversity and greatly increase selection efficiency', and they permit innumerable plants to be reproduced asexually from single cells or small pieces of tissue. The particular technique known as 'callus culture' has been used for many years to clone highly valued horticultural plants such as orchids, and cloning of cuttings is widely practised for tree crops such as tea and palm oil as well as by millions of gardeners for garden plants. According to Longworth 'cell culture' of single cells has unexpectedly, and so far inexplicably, resulted in plants with different properties being re-generated from the same clump of parent tissue. Sugar cane, maize and potato plants regenerated in this way have been found which are resistant to important pathogens. This application of cell culture with the capacity to generate vast numbers of seedlings rapidly has the potential to simplify greatly the hitherto laborious procedures of plant breeding and selection.

However, in terms of current and immediate commercial importance, it is through micro-organisms that biotechnology has its greatest impact. Microbial fermentation processes have been of great commercial significance for centuries, for example in bread, wine and cheese production, and there is the prospect of considerable development. Two recent examples of new processes indicate the sorts of impact that such developments can have. In the late 1960s genetically engineered bacteria were developed which were able to digest corn starch to produce high-fructose corn syrup (HFCS) and which left as a residue corn gluten, which is now an important protein feed for livestock. HFCS has made considerable inroads in the United States and in Japan. It has largely replaced sugar as a sweetener in Coca Cola as well as many other food and drink products. This has helped depress sugar prices to Third World growers. In the European Community steps have been taken to prevent HFCS and other new powerful sweeteners from being produced and from undermining the market for domestically grown beet sugar, which is supported by the Common Agricultural Policy. A second important application of microbial fermentation has been the production of ethanol from sugar cane in Brazil and from maize in the USA. The commercial viability of these processes is critically dependent upon the price of oil as the main non-renewable source of fuel, and, to date, massive subsidies have been required to maintain the Brazilian and USA ethanol programmes. Longworth (1986: 16) states, however, that a new biotechnology, Sucrotech, is being patented, which will not only reduce the cost of producing ethanol from sugar cane, but will simultaneously produce fructose at a cost which will be competitive with HFCS and may thereby reclaim part of the sweetener market for sugar cane. That such possibilities are in prospect indicates how volatile the future might be; biotechnology has tilted the competitive balance from sugar cane to maize, causing economic pressure and even disruption to sugar-cane-dependent economies and may in future switch it back again.

In the long run the capacity to ferment fuels microbially from renewable agricultural feedstocks points to an important long-term reorientation of agriculture if non-renewable oil becomes uncompetitively expensive as a fuel for cars and feedstock for certain chemicals. It suggests that eventually there will be an increased emphasis on agricultural production of industrial feedstocks at the same time as continually increasing food output will be required to feed the expanding world population. All this will require considerable increases in agricultural productivity, to which biotechnology will increasingly contribute, but it will at the same time impose great strains on the natural environment and the structure of agriculture.

PRIVATE-SECTOR CONTROL OF BIOTECHNOLOGY DEVELOPMENT

It has been the role of the public sector to undertake research and development of agricultural technology as a public service to firms which might profit from translating that R & D into a commercial product or process, to farmers who might profit from adoption of the technology, and perhaps most importantly to consumers at home and abroad who benefited from the lower prices resulting from greater abundance. This was particularly true of the phase of change dominated by improvements in plant and animal breeding, the 'old' biotechnology. While the public sector had a role in basic research for chemical and mechanical technologies, the benefits of research expenditure in these areas were easier to capture by private companies investing in R & D, so that increasingly the public sector has taken a smaller role in these areas although maintaining a strong regulatory role in regard to agricultural chemicals in particular. Where it is impossible to prevent others from escaping payment for the research costs, either because the product is easily copied (seeds which can be regenerated by farmers or other firms) or because proposals for new methods can be readily implemented, private firms are understandably unwilling to invest. Machines, insecticides, fungicides and other manufactured inputs do lend themselves more readily to private exploitation. Nevertheless the returns to R & D in these products do depend upon the degree of difficulty potential competitors would have in copying the product. In some cases there are inherent technical difficulties in copying the process, or it would be prohibitively expensive, but in other cases it is the ability to obtain patent protection which creates legal barriers to potential competitors' ability to become 'free-riders' and which protects incentives to private investment in R & D.

Traditionally, however, it has been impossible for plant and animal breeders to obtain patent rights for their products, which is one reason why public-sector R & D has remained so important in this area. For, as under the European Patent Convention (EPC) of 1973, it has been judged impossible for plant varieties to satisfy one of the key criteria to qualify for a patent, namely proof of 'an inventive step'. The application of standard breeding practices to generate new varieties by crossing existing plants or animal strains has not been deemed to be invention. Thus under the EPC one set of exclusions from patentability is 'Plant or animal varieties or essentially biological processes for microbiological processes or the products thereof.'[2]

In the absence of patent rights plant-breeding firms in particular have worked hard to obtain other means of protection and royalties for their products. The history of the development of Plant Breeders' Rights (PBR) is presented by Mooney (1983), and is of particular interest because of the concerns he expresses about the consequences of allowing the basic genetic stock, which underpins agriculture and hence our whole society, from becoming private rather than public property. While Mooney's concerns are expressed in relation to 'old' biotechnology varieties of plants, they are of particular importance with respect to the products of the 'new' biotechnology. For bio-engineered products are capable of meeting the inventive-step criterion of patentability, and both micro-organisms and transgenic animals have now been patented in the USA. Because this raises the likelihood of increased private-enterprise control over the most productive agricultural plant and animal genotypes, issues raised with respect to recent developments in the seed industry are briefly reviewed in the next section.

Before examining developments in the seed industry it is worth touching upon the implications of changes in research policy which stress increasing reliance upon private R & D to develop and exploit the new biotechnology and other technologies. In the UK government has decided that it should withdraw from funding what it terms 'near-market research', that is, research beyond the basic phase and which is preparation for commercial exploitation. This policy is based in part upon the arguments that industry should be investing more heavily in R & D, and that research strategies at the near-market stage should be driven by assessments of likely commercial success which can best be made by the firms involved, and will therefore lead to greater efficiency in allocating research funds.

This has led to the closure of a number of government-financed agricultural research institutes, the scaling down of others and the sale of the National Seeds Organization by auction to Unilever; Unilever won against competition with BP and other major public companies.

This deliberate attempt to switch an increasing proportion of agricultural R & D expenditure from the public to the private sector carries with it a number of risks. In the first place there is a controversy about whether there is under-investment in R & D so that the returns to extra expenditure are high, or whether the converse is the case. If there is under-investment, then withdrawal of public support for applied research will exacerbate this, as the private sector will only undertake R & D expenditure on those products and processes from which exclusive benefits can be captured by the investor. Thus in relation to biotechnology in the USA, where again public support for applied biotechnology research is relatively small, Stallman and Schmid (1987) argue that there will be emphasis on technologies which are applied in factory conditions where secrecy and control can be maintained. For technologies which cannot be confined to factories, such as seeds, Stallman and Schmid state that 'Firms are also considering mechanisms which "scramble" the genome of a plant in the second generation', in order to prevent farmers from reproducing seed with the enhanced, engineered characteristics. Clearly such actions are designed to frustrate the maximum spread of benefits from the technology and to maximize private profit for companies investing in research.

Another facet of commercially orientated biotechnology research is that it will aim at the most important crops and livestock products, those produced by the largest farming units, and those which are most heavily subsidized. In the latter case this will worsen the budgetary problems of adjusting agricultural policy in OECD countries. The probable neglect of minor crops, difficult habitats and small farmers means that the public sector will have a defined role in agricultural technology research, but one in which it will be relegated to the second division and where it is unlikely to prove successful in terms of the commercial yardstick of rates of return which is increasingly emphasized by public-research policy.

DEVELOPMENTS IN THE SEED INDUSTRY[3]

Many major food crops upon which we depend originated and were first cultivated in Third World countries rather than in the industrialized countries where they are most productively exploited today and in which they have been improved by traditional plant breeding. The potato originated in South America, wheat in Ethiopia and the Near East, important maize varieties in South America, and rice in Asia. Moreover the so-called Vavilov centres, which contain the greatest variety of living species and which have been the source of much important gene material, are mainly in Third World countries. The wealth of the industrial countries owes much to their past ability to exploit agriculturally crops and animals originating in other countries. In the colonial era the success of British scientists in smuggling rubber plants from Brazil to Sri Lanka, Singapore and Malaysia and the comparable more complex route by which coffee was introduced to Latin America from Ethiopia are instances of colonial powers obtaining significant wealth by exploiting plant material from the Third World.

In one way or another the process by which institutions and firms in developed countries have continued to collect new plant and animal varieties from less-developed countries has continued. But what is still a major source of friction is the procedures by which legislative protection is developed to give companies in the richer countries commercial rights to exclusive exploitation of varieties developed from plants freely collected from other countries, and to extract high rates of profit from the sale of those varieties. Most extreme is the situation whereby under the USA Plant Patent Act of 1930, which covered asexually propagated plants such as certain fruits, flowers, ornamental shrubs and trees, it has been possible for plants discovered and smuggled out of other countries to be given patent protection in the USA. The key to this is that the Act accepted the unfeasibility of requiring proof of an 'inventive-step' for the granting of a patent, and employed the criterion that as far as can be determined the plant variety is new. Since some novel variety found in the rain forests of South America may satisfy this criterion, and the source of the plant can be concealed from the authorities, the theft and smuggling of plants can be said to have been encouraged.

According to the history presented by Mooney, seed companies in developed countries have successfully pressed their case for protection, and have progressively extended Plant Breeders' Rights (PBR) through both national legislation[4] and the Union for the Protection of New Varieties of Plant (UPOV), which is an international agreement among signatory countries to honour a system of varietal rights. In the case of food crops, in the absence of patentability, PBR operates by registering distinct varieties. Pressure has been exerted upon less-developed countries to become signatories of UPOV and thus acknowledge the rights of breeders, most of whom are from developed countries. However, the majority have stalled or rejected membership, and have been fundamentally unhappy about accepting the principle of creating monopoly rights for others over plant material which in some cases originated in their own country.

The whole issue of intellectual property rights has been raised to a prominent political level by the USA's insistence that it be included as one of the thirteen areas for negotiation in the Uruguay Round of multilateral bargaining of the General Agreement on Trade and Tariffs. While the issue is a much broader one than that of plant and animal breeders' rights, it reflects a basic political conflict between the richer countries, determined to obtain secure returns on R & D expenditures by companies, and LDCs struggling to catch up and anxious to avoid having to pay (excessive[5]) royalties on products protected by PBR and patents. This conflict has broader moral and ethical dimensions when it relates to the issue of charging royalties for the seeds of food crops to countries where malnutrition is widespread.

Against the previous argument has to be balanced the need of those incurring R & D expenditure to capture enough of the returns to provide an incentive for R & D. At least that is certainly the case where R & D is to be undertaken by the private sector, which is politically the increasingly preferred solution in the UK, the USA and elsewhere. This, however, leads to another area of concern, which is that a small number of large multinational chemical and pharmaceutical firms have increasingly come to control the seed industry. Companies such as Shell, Sandoz, Dekallb-Pfizer, Ciba-Geigy and BP have become owners of most of what were small independent seed companies. This linking between chemicals and seeds within firms has a strong commercial logic. It facilitates the marketing of packages consisting of seed varieties plus the appropriate fertilizers, insecticides, etc. It also allows the companies concerned to integrate their plant breeding, biotechnology and chemical research. One particular synergy for these companies will be to bioengineer varieties of major food crops to resist their own herbicides and weedkillers. This would permit chemical weed control to be extended to large-scale crops where this is not now possible, and is just one way in which the direction of biotechnological research may be influenced (biased) by the combination of forces giving a small number of companies extensive control over the basic means of agricultural production.

There are many other issues which could be touched on regarding corporate influence over the development of agricultural biotechnology. It is, however, understandable that the spokesmen for LDC countries, which are profoundly concerned about their dependence upon western technology and exercised by what is seen as colonial and post-colonial exploitation by the West, should be alarmed by the prospect of new rights being created over seeds and other biotechnological products. Certainly there are widespread fears that they will only gain access to the fruits of biotechnology on unfavourable terms and that their relative dependence on and subservience to western industry will be increased so that benefits to them will be small. There are many western critics who would agree with this, and that the system by which agricultural biotechnology is delivered will favour the 'haves' rather than the 'have-nots'.[6]

CONCLUSIONS – BROAD ISSUES RELATING TO THE FUTURE OF AGRICULTURAL BIOTECHNOLOGY

There are divergent views about the rate of uptake of new agricultural biotechnologies and their impact. Kalter and Tauer (1987) and the US Office of Technology Assessment (1986) anticipate comparatively rapid take-up, while others (Buckwell and Moxey (1990) and Farrington (1988)) do not foresee much impact until the next century, that is until ten to fifteen years have elapsed. The caution of the latter commentators is probably justified, for there are various hurdles to be jumped before biotechnologies can gain acceptance.

One hurdle to be overcome is the economic one. Self-evidently there must be some economic advantage to farmers or agro-industry from adopting the technology. Frequently what captures the scientific imagination turns out to be economically unviable in use, although it is only with use that efficiency can be improved and costs brought down.

A more demanding test still for biotechnologies will be to obtain legislative approval and public acceptance. These are intimately connected. In the case of machinery developments, provided safety standards are observed there are no legal impediments to developing new and improved machines, since there are no obvious problems of hazard to the public and therefore public acceptance. Chemical insecticide, fungicides and herbicides do, however, pose much greater problems because of concerns with toxicity, and elaborate testing and licensing procedures have evolved after a relatively haphazard set of procedures in the 1950s and 1960s, when widespread use was made of DDT and organo-phosphorous compounds without adequate recognition of their toxicity. In significant part, because of the problems which have been experienced with agrochemicals, the licensing of biotechnologies will inevitably be based on testing at least as stringent as that which now exists for chemicals. Indeed the licensing procedures are likely to be more stringent because of public and scientific concerns.

Public acceptance of biotechnology in the food chain will be made more difficult by confusion about differences between biotechnologies. In the early 1980s there was a crescendo of concern about the use of steroids to promote faster liveweight gain in calves and cattle. The public outcry eventually led to the banning of such hormones for meat animals; but the legacy is a profound distrust of and hostility to new products such as synthetic Bovine Somatrophin (BST), which has the capacity to increase the milk yield of those dairy cows which are 'relatively' deficient in what is a naturally occurring hormone. Although initial scientific results are favourable, and some minor doubts remain, the major influence leading the European Commission to impose a moratorium on the use of BST until the end of 1990 at least is concern about public perception; the Commission has stated 'It would be a serious setback to producers and to the Community's milk policy were present [positive] trends in consumption to be reversed as a result of adverse consumer reaction.'[7]

With newly bioengineered plants and animals scientific and public concerns are emerging which are of a different order from those associated with previous biotechnology. One relates to the possibility that crop failures may become more frequent if biotechnology leads to a reduction in genetic diversity in crops being grown; such a tendency has already been observed in relation to the Green Revolution technology for wheat. There are also concerns about upsetting the balance of nature in unforseen ways, akin to the unanticipated consequence of introducing rabbits into Australia and then trying to control these by introducing myxomatosis. Thus questions arise such as what happens if herbicide resistance introduced into a commercial crop plant transfers itself by cross pollination into a closely related weed species, or indeed if the herbicide-resistant crop should colonize wild habitats. There are, of course, more lurid and absurd notions bandied about in the popular press which are similar in nature to the attacks made on Darwinism.

The questions, both absurd and real, raised in relation to agricultural biotechnology will undoubtedly slow the rate at which it is adopted. Nevertheless instances of adoption are increasing: BST in the USA, a bioengineered baker's yeast in the UK, bioengineered sheep producing insulin, etc. As these become more widespread and increasingly affect lower-valued, bulk agricultural products, so the impacts of biotechnology on structural change will increase. The balance of economic power will switch increasingly to industry, to high-technology large farms and against smaller farmers in disadvantaged regions and countries. That, unfortunately, is a seemingly inevitable consequence of what we consider to be economic progress.

NOTES

1 In the UK these commercial applications are managed by Protein Pharmaceuticals, a company formed by the AFRC/IRC Centre for Transgenic Animals.
2 For a comprehensive review of patent protection for biotechnologies see Beier *et al.* 1985.
3 This section draws heavily upon the excellent monograph by Mooney 1983.
4 National legislation may confer Inventor's Certificates rather than adopting the PBR forms of legislation, while a number of countries confer patents for asexually propagated plants.
5 According to Mooney many countries are considering a thirty-year patent for plants; and while the North has expressed willingness to reduce protection on industrial patents to five years for southern countries, UPOV makes no exceptions to its long period of protection.
6 See, for example, Goodman *et al.* 1987; George 1979.
7 Quoted in *Green Europe* (October 1989).

REFERENCES

Beir, F. K., Crespi, R. S. and Straus, J. (1985) *Biotechnology and Patent Protection: An International Review*, Paris: OECD.

Buckwell, A. and Moxey, A. (1990) 'Biotechnology and agriculture', *Food Policy* 15(1): 44–56.

Bye, P. (1987) 'Biotechnology and agricultural food complexes', paper presented at International Seminar on Social Dimensions of Biotechnology: Towards a European Policy, Dublin, November 1987.

Farrington, J. (1988) 'A review of developments in agricultural technology and potential implications for the Third World', Paper for an Overseas Development Institute Seminar, London.

George, S. (1979) *Feeding the Few: Corporate Control of Food*, Washington, D.C. and Amsterdam: Institute for Policy Studies.

Goodman, D., Sorj, B. and Wilkinson, J. (1987) *From Farming to Biotechnology: A Theory of Agro-Industrial Development*, Oxford: Basil Blackwell.

Kalter, R. J. and Tauer, L. W. (1987) 'Potential impacts of agricultural biotechnology', *American Journal of Agricultural Economics* 69(2): 420–5.

Longworth, J. W. (1986) *Biotechnology and Agriculture*, Agricultural Economics Papers 1/86, University of Queensland.

Mooney, P. R. (1983) 'The law of the seed: another development in plant genetic resources', in *Development Dialogue* 1–2, Uppsala: Dag Hammarskjöld Foundation.

Office of Technology Assessment (1986) *Technology, Public Policy and the Changing Structure of American Agriculture*, Washington, D.C.: Congress of the United States.

Stallman, J. I. and Schmid, A. A. (1987) 'Property rights in plants: implications for biotechnology research and extension', *American Journal of Agricultural Economics* 69(2): 432–7.

3

GENETIC ENGINEERING AND THE NORTH–SOUTH DIVIDE

Søren Holm

> But at the end of the day, whom will all this activity help? Some people certainly, but, I suspect, they will only be that minority already well supplied with medical goods and services, and the costs of providing cures for the ills of this social stratum will not fall. Too much money is made from putting molecules into people at a profit for very much to change without major convulsions.
>
> That, for me, tarnishes the gleaming image of biotechnology. It is not that most people in health-care companies don't care that their work never touches the lives of millions of people. It is just that they have different priorities: an adequate return on capital invested. Their view of medicine is based on the view that people must first accumulate wealth and then buy health; otherwise there is no deal. The public good may take too long to be a worthwhile business proposition. 'Pure' research is used to fill that gap, but it is being increasingly subject to commercial criteria.
>
> (Yoxen 1983: 141–2)

MAKING BETTER PEOPLE

Genetic engineering in humans holds out large promises for the future, but like all scientific promises these have to be evaluated to see whether it is worth waiting and working for their fulfilment.

For once philosophers and ethicists may be able to survey the 'geography' and the 'maps' before the journey starts, instead of being relegated to their customary role as commentators, discussing what 'route' should have been chosen at points long passed.

Although we cannot predict the exact future of genetic engineering in humans, we have good indications of what the possibilities might be, and this enables us to perform an ethical evaluation before the actual experiments are carried out.

Much of the discussion up to this point has been centred on two major points: (1) Is it at all ethically permissible to alter the genetic make-up of human beings? and (2) Is there a significant ethical difference between interventions that change the human germ line and interventions that only affect the genetic material in somatic cells, but do not touch the germ line (President's Commission 1982; Engelhardt 1990; Hoose 1990)? These issues are not the subject of the present paper. It sets the scene a little further in the future, where it has been decided that society will allow genetic engineering and after the development of successful methods for the performance of genetic engineering in humans.

WHY GENETIC ENGINEERING?

Genetic engineering can be performed for two distinct reasons. What is usually envisioned and discussed is what could be called remedial genetic engineering, that is, the situation where new genetic material is introduced in order to correct some pre-existing defect in the genetic make-up of the individual. This could, for instance, be the introduction of new haemoglobin genes in persons with sickle-cell disease, new hypoxanthine-guanine-phosphoribosyltransferase genes in persons with Lesch-Nyhan syndrome, or, as already has been done, the insertion of new Adenosinedeaminase genes into the lymphocytes and stemcells of children with Severe Combined Immunodeficiency (Anderson 1991). This form of genetic engineering is very much like other medical interventions, both in scope and in intention.

The other and quite different form of genetic engineering is what could be called enhancing genetic engineering. In this form of genetic engineering, the aim is not to correct a pre-existing genetic defect but to introduce genetic material which will make the person 'better' than the usual human norm. It could be genes that would elevate intelligence or enhance height or physical strength, or genes that would endow the individual with 'natural' resistance towards infectious diseases, either generally or against some specific diseases (i.e. a form of genetic vaccination).

Enhancing genetic engineering could, of course, also be used for more frivolous purposes, like producing children with a certain physical look (red hair, brown eyes, six fingers, etc.). If perfected this could be the 'facelift' of the twenty-first century. Such 'cosmetic' use raises separate and rather interesting problems, which, although important, will not be considered in this paper.

There is no sharp dividing-line between remedial and enhancing engineering. What, for instance, is to be said about genetic engineering for greater height in families with familial short stature? But even if the dividing-line is not distinct we can nevertheless separate a cluster of genetic engineering enterprises that clearly fall in the enhancing category.

The perfection and use of this second enhancing form of genetic engineering lie further in the future than remedial engineering, but it is only the enhancing form which has the potential to alter the life of large numbers of people, and thereby the potential to alter significantly society as such.

We do not know whether enhancing genetic engineering for more complex attributes like intelligence or resistance to infections will ever be possible. If it becomes possible, it will, however, create such strong incentives for use that its introduction will be almost irresistible. Society will therefore be well advised to consider the problems that this might cause now in advance, when it is still possible to keep a clear head. Researchers are often hostile towards the prospect of societal control (Lee 1991), but this may be one of the cases where societal control is nevertheless warranted.

THE PROBLEM

One of the largest questions raised by the possibility of enhancing genetic engineering is whether the use of this kind of genetic engineering must be questioned, not because it is dangerous, degrading to the dignity of the human race, or in other ways directly unethical, but because it will lead to the creation of serious and permanent injustice. There are many good things to say about enhancing engineering. These good things are what makes the prospect of a rapid introduction of the technique so tantalizing. The main focus of this paper is, however, on the distributional aspects of the technique; the evident positive aspects will thus be somewhat neglected.

If genetic engineering is used to produce 'better' people (in the non-moral sense), then it is of very great importance how the benefits of such engineering are distributed. If enhancing genetic engineering will widen the already existing differentials in health status between different social classes or broaden the global North–South divide, then there might be reasons to reject it on the grounds of justice.

In order to predict whether such a development is likely it is necessary to make a small technical sojourn, since the distribution of enhancing engineering will depend primarily on its price and on the extent to which technical skills and laboratory facilities are necessary.

GENES AND CELLS

The insertion of new pieces of genetic material into the cells of the human body can in principle be achieved in several ways. The genetic material can be inserted into the fertilized egg or early embryo during the process of in vitro fertilization, it can be inserted into cells temporarily removed from the body (for example bone marrow cells or living skin cells), or it can be inserted through the use of a viral vector or of a liposome. In the viral method the new pieces of genetic material are loaded into the empty shell of a virus, which naturally infects the type of cells in which the genetic material is to be inserted. The virus shell still recognizes the surface receptors of its natural target cell, it connects with it, and it injects the genetic material into the cell (Anderson 1992).

It is obvious that these three methods demand very different levels of technical facilities in the places where the actual insertion of the genetic material is to take place. The first method can only be used in conjunction with in vitro fertilization, and the second demands the ability to extract the relevant cell type and a small laboratory where the genetic material can be inserted into the cells, whereas the third method, at the point of delivery, only demands the ability to inject the loaded viral vector into the person whose genetic constitution is to be changed.[1]

If the technique using viral vectors can be perfected it opens the possibility of performing genetic engineering on a large scale (Johnston 1991), since this technique is not potentially more complicated than a traditional vaccination. The genetic changes made through the injection of a properly loaded viral vector will, however, only benefit the actual person who is receiving the injection. It will not be passed on to any offspring, since the genetic material in the germ cells is not affected.

WHERE THE NEED IS GREATEST?

If we compare the situation of the populations in the First and the Third World, it is fairly obvious that the need for enhancing genetic engineering is much larger in the Third World. Third World populations live in an environment where infectious diseases are still a common cause of disability and death, where industrial pollution is uncontrolled, and overcrowding, poverty and hunger the order of the day. In such an environment genetically engineered resistance against one or more infectious diseases or against environmental hazards would confer a distinct advantage. Engineering for higher intelligence could offset the damaging effects of malnutrition early in life, and engineering for larger muscular strength would still be a benefit in a society relying on human 'manpower' for many projects.[2]

It is, however, equally obvious that enhancing genetic engineering is not going to be widely applied where the need is greatest. If we look to the traditional vaccination programmes against tetanus, poliomyelitis, measles, etc. we see great differences in the vaccination rate in developed and developing countries. It is true that the vaccination programmes in the developing countries must be described as a success compared with many other health programmes, but this does not make the difference disappear between the average 95–8 per cent vaccination rate in the developed countries and the 40–80 per cent in the developing countries, especially since the rates in the poorest countries like Bangladesh and Ethiopia barely reach 25 per cent (Hinman and Orenstein 1990; Hall *et al.* 1990).

Vaccination involves fairly simple technology. Logistically it requires a central producer of vaccines, an unbroken cooling chain between producer and health centre, and a series of health centres dispersed in such a way that they are able to reach most of the population. Central vaccine laboratories have been established in more than twenty developing countries, the cooling chain is kept going with kerosene or solar-powered refrigerators, and widely dispersed health centres are the centrepiece of health policy in almost all developing countries. However, despite the relatively simple technology, the low cost compared to other health-care interventions and the political support for vaccination programmes, it has still not been possible to raise the vaccination rates in the developing countries to acceptable levels (Morley 1989).

It is unlikely that genetic engineering will ever become technologically easier to perform than present-day vaccinations. The vaccination rates must therefore be taken as the upper limit for the possible penetration of genetic engineering into different societies. One could, of course, hope for a major reordering of the structure of world economy and politics that would allow more resources to be put into health care in the Third World, but the developments at the Earth Summit in Rio de Janeiro in the summer of 1992, where the rich countries would not even promise to bring their development aid up to the level of 0.7 per cent of their GNP before the year 2000, does not support great expectations for an economic reordering.

A way to combat these problems could be the use of germ-line engineering, where enhancements are perpetuated in the descendants of the person who had his or her cells engineered. Germ-line engineering seems, however, to be much more technologically complicated and would probably require the engineering to be performed on very early zygotes during IVF procedures. It is therefore unlikely that this technology will ever gain any widespread use in the developing world, where basic maternity care is still lacking in many places.

A very efficient method of getting around this impasse was described as early as 1932, but it still seems ethically repugnant:

> 'Can you tell us the record for a single ovary, Mr. Foster?'
>
> 'Sixteen thousand and twelve in this Centre', Mr. Foster replied without hesitation . . .
>
> 'Sixteen thousand and twelve; in one hundred and eighty-nine batches of identicals. But of course they've done much better', he rattled on, 'in some of the tropical Centres. Singapore has often produced over sixteen thousand five hundred; and Mombasa has actually touched the seventeen thousand mark. But then they have unfair advantages. You should see the way a negro ovary responds to pituitary! It's quite astonishing, when you're used to working with European material . . .
>
> Still we mean to beat them if we can. I'm working on a wonderful Delta-Minus ovary at this moment. Only just eighteen months old. Over twelve thousand seven hundred children already, either decanted or in embryo. And still going strong. We'll beat them yet.'
>
> (Huxley 1983: 5–6)

Although the patent status of pieces of human genetic material is still not quite certain, the development seems to be towards a legal right to patent commercially useful segments of the human genome (Carey and Crawley 1990). If this becomes international law it is predictable that all genetic sequences containing information useful for enhancing engineering will be patented and thereby become the property of some individual or firm. Many of the patents underlying the present biotechnological industry are owned by universities, but the actual development of marketable products is always carried out by commercial firms. The patents would cover not only sequences coding for specific proteins or hormones but also regulatory sequences like tissue-specific initiator or promoter regions.

From a commercial point of view enhancing genetic engineering is far more interesting than remedial engineering, because it is in the enhancing engineering market that large-volume sales will be found. A specific genetic cure for a genetic disease will only be interesting for the few people who actually suffer from this disease, and although prices can be high, the sales volume for each individual 'product' will be very low. The market for enhancing engineering will be quite different. Everybody can use an extra 10 or 20 points in IQ, and given the relationship between social status and height a few inches more would probably be a valuable present for any child. The enhancing engineering market will therefore be a market where both high prices and high volume can be maintained, as long as the product sold is suitably protected by patent. In the market the natural incentive will be towards developing products that bring the highest aggregate projected profits. This relationship between profit and incentive also delineates the class of products for which there is no development incentive.

The concept of 'orphan' drugs comes from the area of classical pharmaceutical products and denotes those drugs used by so few people or by such poor populations that they are not financially viable. What is interesting in the present context is that the lists of orphan drugs contain drugs that are used against widespread diseases like leprosy and river blindness (onchocerciasis) (Anderson 1983; Asbury 1985). What makes these special drugs orphan is not that they are not needed, not that they can only help a limited number of people with a rare disease, but simply that the diseases they can cure are diseases that are prevalent only in the poor developing countries. There is a great need for these drugs, but there is no market, and the only reason that they are actually still produced is either that the firms holding the patents are given tax deductions for their 'community service' or that they are able to use such 'philanthropic' actions in their marketing efforts in the western world.

If we draw an analogy with the field of genetic engineering it is fairly obvious that the largest profits lie in developing products that appeal to a significant segment of the population in the western countries, i.e. by engineering for intelligence, height or beauty, and against atherosclerosis and cancer. Some tropical diseases (for example malaria) pose such risk for westerners that it will probably be economically viable to develop products giving genetic resistance against them. It is more questionable whether engineering for an increased general resistance against infections will be of interest in the western world, and there will definitely not be large sales in genetic engineering products offering specific resistance against leprosy, river blindness, hookworms or sleeping sickness, to mention a few major infectious and parasitic diseases of great importance in the developing countries. It is therefore unlikely that the biotechnological industry will develop any of the products needed primarily in the Third World (Teitelman 1989).

One could hope that such products would instead be developed by national or international public research institutions, but this hope has no grounding in the previous performance of this sector in the development of more traditional drugs.

Of the products that are likely to be developed some of those likely to be most profitable will only be of minor interest in the Third World, where the present lack of even basic health care makes diseases like atherosclerosis and cancer a secondary or tertiary priority. Products aimed at the prevention of such diseases are only of interest in the developed countries, where the average length of life is sufficiently high for these diseases to manifest themselves and the life-style sufficiently 'modern' for them to develop.

DIFFERENCES, EQUITY AND ETHICS

If enhancing genetic engineering can bring large benefits, if the beneficial effects are most needed in the Third World, and if our global social and economic structures make it unlikely that enhancing engineering will be used most where it is most needed, what should we then think of this state of affairs from the point of view of ethics?

We live in societies that accept great social differences in many areas of life. Even in the western world some people go hungry every day, illiteracy is not eradicated, and many are homeless, but these social class-related differences of course pale in comparison with the differences found between the First and the Third World. Although most probably feel a little unease when they watch yet another famine victim on the television screen, we seem to be able to dispel that unease by donating ten pounds to Oxfam or by performing some voluntary work in our own community. If the great present differences do not bother our moral senses that much, why should we then care about these new differences created by the introduction of enhancing genetic engineering?[3]

There are two important features that distinguish the possible effects of enhancing genetic engineering from the already existing differences caused mainly by economic factors.

The first is the relation of these differences to the often-used distinction between nature and nurture. It is perhaps best brought out if we compare genetic engineering of higher intelligence with differences in intellectual prowess brought about by differences in education. In a system where differential schooling is prevalent even in the compulsory parts of the school system (for example England or the world taken as a whole), those who are able to buy into the better schools will no doubt get better education, even though this is just one of the reasons why people are willing to pay to get their children into a specific school. The main reason why a school like Eton is able to demand high fees is not that it is a place of academic excellence. The fees are primarily payment for the social status and the social network that accompany an education at Eton. The academic advantage achieved by better education is, however, not permanent: many of the things we learned at school twenty years ago are now obsolete, and good study habits and techniques are much more a question of personal style than of expensive education. Neither is the advantage freely transferable to new fields of academic study. If an Etonian and a former pupil of some Moss Side school in Manchester (i.e. a state school in a deprived inner-city area) were to compete in the learning of some totally alien subject, say Inuit languages, the money spent at Eton wouldn't be of much benefit. The result of the contest would depend entirely on motivation and intelligence. If, however, the Etonian's parents had instead spent their money on a genetic engineering product raising the IQ of their child by twenty points, it would secure the child and the adult it grows up to be a significant advantage in all intellectual subject areas, even the most alien (whether this could compensate for the lack of the old boys' network built up at Eton is a matter I will leave aside for the moment).

The anthropologists have used the last hundred years to dispel the once-popular belief in the existence of primitive people. They have been so successful that when we today compare ourselves to the people in the Third World we see them as unfortunate victims of social and economic circumstances, not as less worthy people biologically determined to live as our servants/slaves, and with Kipling we might even say:[4]

> Though I've belted you and flayed you,[5]
> By the Livin Gawd that made you,
> You're a better man than I am, Gunga Din.
>
> (Kipling 1981)

This is, however, an attitude that is closely connected with the belief that the reason why the people in the Third World are poor, disease-ridden and illiterate is not that they are in some way genetically or biologically predestined to this fate. If we erode the basis for this belief, we may also erode any public willingness to support development aid. In present European discussions about immigrants from Third World countries a lot of attention is being focused on the culture of the immigrants. It is often conjectured that their culture prevents them from becoming full productive members of society, and special programmes aimed at keeping immigrant children in touch with their own language and culture are losing public support. If a belief in western cultural superiority – albeit on the home ground – can support far-reaching political decisions, what kind of decisions would then flow from a (warranted?) belief in biological/genetic superiority?

An interesting example of the connection between the belief in (biological) predestination, moral worth and resource allocation is found in the South African apartheid system. This system was supported by the Reformed Synod of South Africa, who claimed to have biblical evidence that the dark races were predestined to be the servants of the whites because of the sins of Noah's son Ham.[6] From this belief it was easy to argue for separate education and separate 'rights'.

Some twenty-five years ago there was public uproar when an American scientist claimed to have evidence that genetic constitution was the main determinant of intelligence and that the low social status of blacks in American society showed that they were genetically determined to be less intelligent than whites (Jensen 1969). In the debate that followed both data and conclusions were criticized and discredited (Wade 1976; Blum 1978; Taylor 1980). We may, however, ourselves be able to create a situation in the future where this claim would come true.

Our society values intelligence highly (at least when it is used for something more useful than moral philosophy). In a future situation where many people in the West were genetically superior to people in the Third World in this area there would be a great temptation to revert to the old claim that this would in some sense make us better people than they.

It is, of course, trivially true that a person who is genetically engineered to have a higher intelligence than he would otherwise have is better (in a technical sense) than he would otherwise have been. What is important is that this value judgment only pertains to his intelligence and not to his person as a whole and even less to his moral status. There are, however, many possibilities for equivocation surrounding the word 'better', and the slide from the technical connotation (higher intelligence) to the value connotation (more valuable) is much more likely to occur when the 'better' is connected to an attribute that is perceived as genetically determined than when it is connected to a clearly social attribute (for example wealth). The popular conception of personality is still thoroughly essentialist. I am the person I am because there is some core within me that has determined me to become this person. This core is often identified with the genetic constitution, and it is believed that it is directly inherited from one's ancestors. This view of personality is evident in casual conversations, where statements like 'He has his father's temper', 'She is as cunning as her mother', or 'That family has always had a nervous constitution' are common. If differences in average intelligence between the West and the Third World (or between social classes in a rich society) were caused to fall clearly on the side of 'nature' and not on the side of 'nurture' as we are currently taught to believe, then the doors would be wide open for a (political) equivocation between the technical 'better' and the moral 'better'.

> *Why should we use money for education in the third world when the people there are genetically and intellectually inferior?*

> *Should not those who are (biologically) best equipped to rule (i.e. we in the West who are genetically and intellectually superior) be allowed to rule?*

Although arguments of this kind are fallaciously based on an underlying equivocation, they are nevertheless rhetorically very powerful. Rhetoric is usually more successful in influencing the formulation of public policy than good philosophical analysis, and a development that would make the premises of such rhetoric true could be dangerous. An additional factor that might add even greater force to this dangerous rhetoric is that the global division between those who will have enhancing engineering and those who will not will follow racial lines, and although racism is universally decried, it is far from dead.

The second important feature that distinguishes the differences between populations created by enhancing genetic engineering is that they are *new* differences of a kind we have hitherto not experienced. There may be reasons why we cannot (or perhaps even should not) change the already existing differences between the rich and the poor, but these reasons may not support the deliberate creation of new differences. As is argued above, the introduction of enhancing genetic engineering is likely to lead to new differences between the populations in the rich western countries and those in the poor countries of the Third World. These differences do not just occur as through some natural process: they are the result of a deliberate development of a new kind of technology, and those who support or allow this development cannot escape the moral responsibility for the results of the development.

The theory of justice is plagued by endless discussions about the exact content and formulation of the principles of justice, but it is possible to distinguish two major schools. Either justice is primarily a result of just processes (the historical view), or justice is primarily connected to a just end-state (Nozick 1974; Rawls 1972). The philosophical defence for the present differences can then take three forms:[7]

(a) The present differences were created through just processes and are therefore just.
(b) The present differences are seen *sub specie aeternitatis* as a just end-state either because they are beneficial overall or because they conform to some principle of justice.
(c) The present differences are not just, but any attempt to

rectify this injustice completely will create more injustice than is removed, or a rectification will be so complicated that it is not possible to complete.

No one with even the slightest knowledge of modern history can hold proposition (a), since the distribution of the global wealth is mainly the result of our violent exploitation of the Third World. Neither a teleological nor a deontological point of view lends much support to proposition (b). It is hard to see how the present distribution of wealth and the attending differences in education, health status, perinatal mortality, etc. can be beneficial overall, and a deontological principle holding the distribution to be just will have to be very strange indeed. The argument for upholding the present distribution must therefore be some version of proposition (c).

If this is true, it restricts the possible justifications for introducing new differences. If the present original situation is unjust, then only theories that acknowledge the importance of both process and end-state or theories that are purely end-state-based can be applicable. Pure process theories will be disqualified from the start because even a just process will not lead to justice if the point of departure is decidedly unjust.[8] The strict Nozickian view (Nozick 1974), basing justice not on end-states but on a chain of just transfers between free individuals, is therefore not applicable given the situation from which the process is going to start. But pure end-state views do not fare much better. Even though it matters very much whether a future state of affairs F is more just than the present state of affairs P, it also matters very much how the transition from P to F took place. This is the core of proposition (c), stating that the transition itself can produce such a large burden of injustice that it cannot be outweighed by the desirable end result. It is true that the end may justify the means, but not even the most desirable ends justify all means.

In the move from P to F, we therefore have to pay attention both to the moral 'costs' incurred at every step of the transition process and the moral 'gains' won through the whole transition.

Considering the introduction of enhancing genetic engineering as a possible step in such a transition process we thus have to ask two questions:

(1) How do the two end-states compare morally?
(2) Are any important moral costs incurred in the process?

It is argued above that the introduction of enhancing genetic engineering is likely to lead to a considerable widening of the differences in health and intellectual status between the populations of First and Third World countries. The size of any trickle-down effect is hard to predict, but even a considerable economic investment in genetic enhancement programmes in the Third World could not eliminate the differences. Will the gains in the richer parts of the world then be able to outweigh the widening of the health-gap? Probably not. The certain gains will be some more life years added to already rather long lives, and unless the ultimate breakthrough can be achieved and the inexorable progress of ageing be conquered, the absolute value of this benefit may be questioned. Whether a general 'improvement' in height, strength, or intelligence would be a benefit at all is even more questionable. To the individual such improvements will benefit his or her social status, but only as long as the same improvements are not so widespread in society that most people share them, thereby again levelling the playing field.[9] In general there seems to be no connection between intelligence and happiness, or intelligence and preference satisfaction. That an improvement in average intelligence would increase average (or absolute) happiness is by no means certain. Greater intelligence could, of course, also be a benefit if it led to a better world through more prudent decisions and useful inventions. For this suggestion there is little empirical evidence; the prevailing political structures would probably in all circumstances ensure that political leadership was not attained by those most suitable to govern. It is perhaps worth noting that neither the main problems facing the world nor suitable solutions for these problems are unknown. The main difficulty seems to be a lack of will to do anything decisive to implement the solutions.

An interesting development could occur if enhancing genetic engineering for some reason turned out to be so expensive that only a minor part of the population could afford it, even in the rich countries. In that case the societal disruption and the disenfranchisement and alienation of the poor would occur within First World countries and not between First and Third World countries. Because of the positive feed-back inherent in the process (i.e. better biology – greater income – better biology, etc.) the end result of such a development could be a set of classical two-tier societies like Athens or Rome. Like these classical examples any modern version of such a society would sooner or later crumble because of internal discord.

All in all I wish to suggest that the benefits of enhancing genetic engineering may not be as great as they initially appear, given a scenario where only the commercially viable products are put on the market. This would change if genetic products aimed at Third World problems like infections were developed and made widely available, partly because the number of life years gained would increase substantially. The end-state reached after the introduction of enhancing engineering would therefore be not much better than the original situation before the introduction, and given the possible side-effects outlined above it might actually be worse.

But what about the process getting us from one end-state to another? If the original state was just, there would be some difficulties in faulting the process involved in the introduction of enhancing genetic engineering, since it is only based on the usual exchanges between 'free' economic agents. But given the fact that the original situation is unjust, it must always be asked whether there is not another process that could produce an end-state with the same desirable features and at the same time a partial rectification of the pre-existing injustices. Any such process will evidently have to be based on something other than the 'free' economic processes. Many different possibilities come to mind, ranging from very strict regulatory measures restricting access to enhancing genetic engineering to the more mildly encouraging linking the development and production of genetic products relevant in the Third World to tax breaks or marketing rights. It is unlikely that even a legal linkage between the rights to market profitable products in the western countries and an obligation to develop and produce products for the Third World would discourage the major pharmaceutical firms from entering the potentially highly profitable market for enhancing genetic engineering. Such a linkage can therefore be one possible element of a process that will bring the same or larger benefits as the 'free' economic process without enlarging the already existing injustices.

Introducing enhancing engineering entirely through the 'free' market may in itself have unwanted side-effects. Just as the differences created (for example in intelligence) could change our view concerning those who are not 'improved', the idea inherent in the market approach, i.e. that money can buy everything, even profound changes in our individual human biology, may lead to a further alienation of the non-privileged. Will it not be natural for those not 'improved' to think that the introduction of such deep biological differences strictly correlated to wealth will make their liberation from poverty an impossibility?

CONCLUSION

I have argued that both the end-state brought about and the process involved in an essentially unregulated introduction of enhancing genetic engineering through the market are questionable seen from the point of view of any applicable theory of justice. When this is combined with the attending serious side-effects, the conclusion has to be that only a regulated approach can prevent (or perhaps only ameliorate) great future injustice.

NOTES

1 The method using a viral vector can be used both on cells removed from the body and as described on the whole body through injection of the vector. Present experiments in humans only use the viral vectors on cells removed from the body.

2 We may well ask whether enhancing genetic engineering is the best way to help people in the developing countries, and the answer is probably 'no'. But it is nevertheless a possible way to help them.

3 I here leave aside the interesting question of our present ethical obligations towards the suffering people in the Third World. I believe that we have large duties in this respect that we presently do not discharge, or discharge only to a very minimal extent.

4 I do not wish to imply that Kipling held the view about the connection between nature and nurture that I am describing in the following section.

5 In our case through economic exploitation.

6 See Genesis 9: 20–7.

7 There is a fourth political defence with the form: The present differences are not just, but we who have the power to decide are not willing to try to rectify the injustice.

8 The decision rule 'never perform an act unless the act reduces injustice' does not lead to a pure process theory, since the decision rule relies on a comparison between the two end-states.

9 What would be the status of Eton, Oxford and Cambridge if all could go there?

REFERENCES

Anderson, Kenneth (1983) *Orphan Drugs*, New York: The Linden Press.

Anderson, W. F. (1991) 'End-of-the-year potpourri', *Human Gene Therapy* 2: 299–300.

— (1992) 'Human gene therapy', *Science* 256: 808–13.

Asbury, Carolyn H. (1985) *Orphan Drugs: Medical versus Market Value*, Lexington, Mass.: Lexington Books.

Blum, Jeffrey M. (1978) *Pseudoscience and Mental Ability: The Origins and Fallacies of the IQ Controversy*, New York: Monthly Review Press.

Carey, N. H. and Crawley, P. E. (1990) 'Commercial exploitation of the human genome: what are the problems?' in: *Human Genetic Information: Science, Law and Ethics*, Ciba Foundation Symposium 149, Chichester: John Wiley & Sons (pp. 133–47).

Engelhardt, H. Tristram (1990) 'Human nature technologically revisited', *Social Philosophy & Policy* 8: 180–91.

Hall, A. J., Greenwood, B. M. and Whittle, H. (1990) 'Modern vaccines – practice in developing countries', *Lancet* 335: 774–7.

Hinman, A. R. and Orenstein, W. A. (1990) 'Modern vaccines – immunisation practice in developed countries', *Lancet* 335: 707–10.

Hoose, Bernard (1990) 'Gene therapy: where to draw the line', *Human Gene Therapy* 1: 299–306.

Huxley, Aldous (1983) *Brave New World*, London: Longman Group Ltd.

Jensen, Arthur R. (1969) 'How much can be boost IQ and scholastic achievement?', *Harvard Educational Review* 39: 1–123.

Johnston, Jeff (1991) 'Gene therapy: after the dream world, the serious science', *Journal of NIH Research* 3 (June): 26–9.

Kipling, Rudyard (1981) 'Gunga Din', in David Herbert (ed.), *Everyman's Book of Evergreen Verse*, Everyman's Library, London: Dent.

Lee, Thomas F. (1991) *The Human Genome Project: Cracking the Genetic Code of Life*, New York: Plenum Press.

Morley, David (1989) 'Saving children's lives by vaccination', *BMJ* 299: 1544–5.

Nozick, Robert (1974) *Anarchy, State, and Utopia*, Oxford: Basil Blackwell Ltd.

President's Commission for the Study of Ethical Problems in Medicine and Biomedical and Behavioural Research (1982) *Splicing Life*, Washington, D.C.: US Government Printing Office.

Rawls, John (1972) *A Theory of Justice*, Oxford: Oxford University Press.

Taylor, Howard F. (1980) *The IQ Game: A Methodological Inquiry into the Heredity–Environment Controversy*, New Brunswick, N. J.: Rutgers University Press.

Teitelman, Robert (1989) *Gene Dreams: Wall Street, Academia, and the Rise of Biotechnology*, New York: Basic Books.

Wade, Nicholas (1976) 'IQ and Heredity: suspicion of fraud beclouds classic experiment', *Science* 194: 916–19.

Yoxen, Edward (1983) *The Gene Business: Who Should Control Biotechnology*, London: Pan Books.

4

THE FRUITS OF BODY-BUILDERS' LABOUR

Hillel Steiner

When it comes to moral impediments, the right of self-ownership has to be a leading contender for the league championship. What other moral value stands in the way of as many significant ethical and political objectives as this right does? Whether it be the alleviation of needs or the realization of what has been called 'equality of condition' or the maximization of some highly desirable social variable, sooner or later its redistributive demands seem bound to run into head-on conflict with those of self-ownership.[1]

So why do we continue to allow this vexatious value to obstruct or constrain our pursuit of noble projects? Not all of us do, of course. Nor am I competent to supply a full and instructive answer for the rest of us who do encumber our practical reasoning with this restriction. But there can be little doubt that one part of that answer lies in the fairly entrenched revulsion, unmistakably present in our pre-theoretical moral thinking, against such practices as slavery and exploitation.

For even pre-theoretically, it's pretty clear that a consistent vesting of any *other* rights, in persons whose bodies are owned by others, must remain an unperformable intellectual feat. Slaves' bodies are uncontracted organ banks. Exploited and enslaved persons are deprived of their bodies' labour by being deprived of the products containing it. The labour embodied in those products is itself a product of the labourer's body: production is a transfer of energy from that body to those products. Matter and energy are mutually convertible. And those products thus incorporate part of that body just as they would if portions of that body were to be detached from the rest of it and literally incorporated into those products: a body's mass decreases when its energy is expended without being replenished. Self-ownership is what is needed to render slavery and exploitation non-contingently impermissible. And this holds true for any conception of slavery or exploitation we might favour.

It's one thing, however, to say that categorical moral rejection of these practices logically entails self-ownership. It's quite another to regard this right as easily integrated into general theories of moral rights. For the vast majority of such theories, as Robert Nozick has strikingly demonstrated, tend to generate rights which are incompatible with self-ownership and which thus permit and indeed mandate actions that are in violation of it.[2] So it is perhaps the chief intellectual virtue of Nozick's Historical Entitlement Theory that it seeks to embed this right, so strongly associated with our ordinary views of what is central to justice, in a set of rights compatible with it.

But again, seeking is one thing and finding quite another. And this particular quest is not made any easier by the fact that self-ownership – as a right ascribed universally to all persons – appears to be internally inconsistent and utterly paradoxical. Precisely *why* it is paradoxical, how Historical Entitlement Theory can enable us to resolve this paradox, and what the wider distributive implications of this resolution are constitute the three items forming the agenda of this essay.

1 THE PARADOX OF UNIVERSAL SELF-OWNERSHIP

Excavation of this paradox could do worse than to begin with the *reductio* Herbert Spencer performed on Proudhon's claim that there is no moral right to private property:

> If no one can equitably become the exclusive possessor of any article – or as we say, obtain a right to it, then, amongst other consequences, it follows, that a man can have no right to the things he consumes for food. . . . Wherefore, pursuing the idea, we arrive at the curious conclusion, that as the whole of his bones, muscles, skin &c., have been thus built up from nutriment not belonging to him, a man has no property in his own flesh and blood – can have no valid title to himself – has no more claim to his own limbs than he has to the limbs of another – and has as good a right to his neighbour's body as to his own![3]

Now, while Spencer may have been a trifle hasty in simply presuming this to be a knock-down argument for private property rights – while he may have been over-confident of self-ownership's moral incontestability – he was surely right in his suggestion that the ownership of products bears some close moral dependency on the ownership of factors entering into their production.

Thus, for example, I own the oranges sold to me by my grocer. And I own the orange juice I make from them with my juice-making machine and the electricity I have paid for. If I use that orange juice, along with other ingredients and instruments which I own, to make a cake, then I own that cake. If, instead, I drink the juice, then I own the resultant body tissue which is the joint product of the juice and other bodily ingredients which I own. And I further own one of the main products of that body: namely its labour.

So far, so good. But now it's time for a confession. You see, although it's been true for many years that my body tissues have been produced from factors belonging to me, I feel bound to reveal that this was not always so. I'm not now referring to the glass of orange juice I once stole from the high-school cafeteria. What I have in mind, rather, is that in the earlier period of my life and even before I was too young to know better, the production of my body tissues was conducted largely by persons other than myself with factors that were owned by them and not by me. During that nurturing period it was my parents who supplied my orange juice; prior to that, it was they who provided my ante-natal subsistence; and later it was they who laboured, or hired the labour of others, to perform the often thankless task of enhancing the informational content of my brain, the dexterity of my limbs and my immunity to disease. And while the number and identity of who are to count as my parents might arguably depend upon whether I was conceived in the conventional way or in an IVF clinic or by a parthenogenetic process or in my local neighbourhood clone factory, the general point is that it was they and not I who built the body to the production of which I only eventually began to contribute.

Do these revelations impugn my self-ownership? Sir Robert Filmer thought they did.[4] According to him, the historically earliest human parents – or as he would have it, Adam – owned all their offspring as the fruits of their labour. And these titles have descended down a primogenital line to the present day. So we are all the property, the slaves, of the current successor of those primordial ancestors. And he is the only self-owner in town. Hence Filmer's advocacy of absolute monarchy.[5]

This is not good news for those of us who oppose exploitation and have pretensions to self-ownership. And what's more, the story doesn't get a lot better even if we tinker around with some of the empirical premises of Filmer's argument. For suppose that, in the light of subsequent historical and scientific evidence, we postulate a somewhat greater plurality of primordial ancestors than the Adamic singularity dictated by Filmer's scriptural convictions. Let's further suppose that not all of them and/or their heirs invariably shared a commitment to singular, let alone primogenital, succession. Indeed, let's go even further and optimistically suppose that, over long stretches of historical time, many of the more liberal of these successors chose – or would have chosen, if they had not been expropriated – to emancipate those whom they owned. Can today's aspiring self-owners take much comfort from such occasional outbreaks of liberalism, actual and conjectural?

I think the answer must be 'no'. For the most we can say, on the basis of this modification of the descriptive elements in Filmer's tale, is that current self-owners would constitute a larger proportion of the total human population than he can allow. There could, however, still be a set of persons, perhaps the vast majority, who would be the property of others. Specifically, they would be owned by their oldest living and self-owning non-liberal ancestors or by other persons to whom those titles had been assigned by their previous owners. And they would be owned well beyond the age at which I started supplying my own orange juice. If we make a person who then makes widgets, how is this different from our making a machine which then makes widgets? Aren't our grandson's widgets as much ours as the widgets churned out by our machine which was itself made by our other machine for making widget machines? Or, if not ours, don't they all belong to whoever owns us?

The outlines of a paradox now emerge more sharply, and painfully. On the one hand, we want to be able to say that everyone is necessarily a self-owner and, thus, that these titles are not ones contingently bestowed because of the emancipatory inclinations of primordial ancestors and their successive assignees. On the other hand, such an affirmation – by denying primordial ancestors the ownership of their offspring – appears thereby to impugn *their* self-ownership by denying their titles to the nurturing factors that produced those offspring, including their own labour. So it looks as though either these ancestors *are* self-owners: in which case most of their descendants, i.e. most of us, needn't be; or these ancestors are *not* self-owners: in which case, what grounds can there possibly be for ascribing self-ownership to anyone else? The title-search for universal self-ownership is clearly a difficult quest. How do we do it?

2 THE PARADOX RESOLVED

We do it, I think, by availing ourselves of the normative resources furnished by the Historical Entitlement Theory of distributive justice. On the face of it, however, this looks to be a singularly unpromising strategy. It's unpromising because Filmer himself might well register a strong claim to being an historical entitlement theorist. For when it comes to the normative as distinct from the empirical details of his case, the distance between him on the one hand, and Locke and Nozick on the other, narrows dramatically.

True, Filmer emphatically denies what both Locke and Nozick affirm: that

> every Man has a *Property* in his own *Person*.[6]

But Filmer's challenge to them and to us is to take seriously and consistently the inference which Locke himself immediately draws from this claim only two sentences later when he says that

> The *Labour* of his Body, and the *Work* of his Hands, we may say, are properly his.[7]

If we are all the products of the labour of other bodies and the work of other hands, how can anyone other than primordial ancestors have an unbestowed property in his or her person? Or conversely, how can we meet Filmer's challenge and expel him from the Historical Entitlement Theory club?

Historical Entitlement Theory postulates four distinct ways in which we can come to own things, four rules for generating titles:[8]

1 by *transfer* – by others transferring to us the ownership of things (including labour) they own, whether by gift or sale or other contractual arrangements;
2 by *appropriation* – by our being the first to acquire from nature some unowned thing (subject to certain distributive provisos which will be discussed presently);

3 by *rectification* – by our acquiring ownership of things held by others who have violated our rights and consequently owe us those things in compensation;
4 by what I shall call, quite generally, *begetting* – by our making or producing something from/with things we already own.

Now, consonant with Historical Entitlement Theory, the position we want to avoid contends that the title to a current live human body is found by searching for the titles to the factors entering into the production or begetting of that body. But that position diverges from our casual and unreflective reception of that theory by insisting that, while such titles *might* be vested in the person whose body it empirically is, they are so only if those who previously owned that body granted that person emancipation by transferring its ownership to him or her. Those previous owners might, but need not, have been that person's parents.

Indeed, they could *not* have been his or her parents if those parents were not themselves self-owned. Faithfully applying the transfer and begetting rules, the position we are considering validates only that title to a person's body that possesses an ownership pedigree consisting – like the pedigree of any other valid title – in an unbroken series of antecedently valid titles, each successively created by the immediately previous title-holder's acts of begetting or title-transfer. Hence all such pedigrees must contain and originally terminate in the title of a primordial ancestor.

I'm going to call this position the Primordial Ancestor Pedigree position, or PAP for short. Both the singular primordiality of Filmer's account and the pluralistic primordiality of modern biology are consistent with PAP. How, then, can we impugn PAP's historical entitlement credentials?

My suggestion is that this can be done by taking a closer audit of the factors entering into the production of one offspring of some primordial ancestors. Let's call him Cain. Cain's body tissues, like ours, were the products of orange juice and chocolate biscuits. Equally, his tissues were moulded by the education, exercise and medicine he received and, before that, by the ante-natal environment supplied to him. All these factors of production were, so to speak, bought and paid for by his parents. They were undeniably his parents' labour and the fruits of their labour.

In order to sustain the claim that – notwithstanding his parents' ownership of these production factors – Cain owns his body tissues, we need to be able to construe their supplying of these factors as amounting to their *divesting* themselves of the titles to them. One way in which people divest themselves of the titles to things they own is by choosing to donate them as gifts to others. But although labourers donating their products is neither exploitative nor enslaving of them and in no way impugns their self-ownership, this route for reaching our desired divestiture conclusion is not open to us.

Why not? Essentially because postulating divestiture by donation is otiose in this context. We are looking for premises that sustain Cain's self-ownership even in the absence of an act of emancipation. Emancipation consists in the owners of Cain's body choosing to donate their title to his body to him. But we don't want Cain's self-ownership to be predicated on the choices of others. So there is no relevant difference between grounding it in divestiture by donation and grounding it in emancipation. Periodic festivals of emancipation and biblical jubilees are certainly to be applauded in the absence of anything better. But they won't do for our purposes, because they are not consistent with universal self-ownership as a matter of right and *not* a matter of other persons' discretionary bestowals.

There is, however, another route for reaching our desired divestiture conclusion, and one better suited to our present purpose. Under Historical Entitlement Theory's begetting rule, it is *ceteris paribus* true that products of labour rightfully belong to those who owned the labour and other ingredients with which it was combined to make those products. But they have to have owned *all* those ingredients. Mixing their labour and other belongings with something that doesn't belong to them does not necessarily give them the title to the thing thereby produced. My working night and day and employing only my own materials to restore an antique chair which I've stolen is not going to give me title to the chair when the job is done. Rather than gaining a restored chair, I've divested myself of my labour and materials. Just as in Locke, I lose or fail to acquire title to a piece of land – and thereby divest myself of the labour and materials I invested in its cultivation – if I can neither consume nor otherwise use its crops.[9]

How does this mode of divestiture, which we might call 'supervenient divestiture', save us from being pinned for ever on the horns of our dilemma? How does it save us from having to deny either Cain's self-ownership or that of his parents? It can serve to do so only if, among the essential production factors of Cain's body, there is at least one item which cannot be said to belong to his parents.

And there is indeed just such an item. Our audit of production factors was seriously incomplete. For apart from the orange juice, chocolate biscuits and ante-natal environment owned and supplied by Cain's parents, there is a set of other ingredients that entered into this body-building process and that are *not* the fruits of his parents' labour. They are not the fruits of his primordial parents' labour because, as we know from Darwin and molecular biology, they existed prior to his parents and certainly prior to their labours. I refer here, of course, to Cain's *genes*. Cain's genes are natural resources because, as Darwin tells us, Cain's *grand*parents were natural resources.

Now, under Historical Entitlement Theory, things are either the products of labour or they are natural resources. And while it is the begetting rule that governs ownership of labour products – assigning it to the owners of that labour – it is the *appropriation rule* that governs the ownership of natural resources. The appropriation rule, as I mentioned earlier, confers the ownership of unowned things on those who first acquire them from nature. But unlike the ownership of labour products, the ownership of natural resources is encumbered. Natural resources, not being products of anyone's labour, are presumed to be things to the use or value of which all persons have an equal claim. Hence any person's appropriation of a natural resource occasions a duty in him or her to compensate others.

Questions of what form this compensation should take and how to construe it as owed, not only to the appropriator's contemporaries but also by his or her successors to members of subsequent generations, have long occupied the attention of philosophers and economists interested in the moral basis of property rights.[10] And although diverse solutions have been proposed, they share in common the view that a natural resource can be privately owned only so long as due compensation continues to be paid by its owners to its non-owners. Its non-owners have the right to that compensation.

This requirement does, I think, provide us with a basis for extricating ourselves from the paradox of universal self-ownership. Universal self-ownership means the self-ownership of all moral agents. So the appropriation rule's compensation requirement vests rights to this compensation in all moral agents. It follows that Cain's parents cannot continue to satisfy this requirement once Cain attains moral agency. For someone who is the property of others cannot be owed compensation or anything else since, as a slave, he lacks all rights and whatever he possesses is the property of his owners. Hence Cain's parents can be said to own him up to the time of his attaining moral agency but not beyond that. At that point, they may be said to be superveniently divested of that ownership. Nor does his self-ownership beyond that point impugn their own self-ownership, because it is logically impossible for the condition of their continued ownership of him – namely that he as well as others are entitled to compensation – to be satisfied. Universal self-ownership, is thus sustained by the encumbrance imposed by the appropriation rule's proviso.

Before examining some of the policy implications of this analysis, I want briefly to review and refine my claim that Cain's genes are natural resources and not the products of his parents' labour. As I suggested previously, a schematic rendering of Darwin's thesis is that although Cain's parents – as primordial humans – are moral agents, his grandparents are not. They, like members of other species, count as natural resources. Being natural resources, they are themselves eligible for appropriation by moral agents: they can become objects of property rights. Suppose they are appropriated by Cain's parents. And further suppose that, in due course, these owned grandparents produce another set of human offspring – Cain's aunts and uncles.

The question we now need to answer is this: do Cain's parents become the owners of these, their siblings, just as they become the owners of the offspring produced by any of their other livestock? Do their siblings become their slaves? I think the answer must clearly be 'no'. Whatever reasons we have for regarding Cain's parents as having attained moral agency at maturity – and hence self-ownership – are going to be reasons for according the same status to Cain's aunts and uncles when they have matured.

What is the difference between the process by which Cain's aunts and uncles were produced and the process by which Cain himself was produced? Suppose that Cain's parents are lonely and would like some more human company. Further suppose that although they are unable to sustain another pregnancy, they are both extremely talented molecular biologists who are fully equipped to perform the relevant kinds of microsurgery with successful outcomes. Whether they choose to operate on their parents or themselves or some combination thereof, it seems clear that they can be described as tapping into one and the same thing: namely the *germ line* shared by all of them.

Whether they produce a sibling for Cain or for themselves or some Oedipal combination of the two or some other relative altogether, they must mix their labour and belongings with an ingredient supplied by nature. *They must appropriate before they can beget*. This is as true of the production of Cain as it is of the production of Cain's aunts, uncles, siblings or whomever. Nor would the fact that production of these persons might be initiated by more conventional processes make any relevant difference to the priority of appropriating over begetting. It's because the production of any person entails a prior appropriation of natural resources that the appropriation rule applies to such productive activities. And that rule imposes a duty on the appropriators to compensate all others.

3 POLICY IMPLICATIONS

This account of universal self-ownership, of how both begettors and begotten can own themselves under the rules of Historical Entitlement Theory, has a number of interesting policy implications. Thus while universal self-ownership is surely good news for all of us, it is not such good news for those who would like also to claim ownership and control of their grown children.

Nor will the route we have followed, to Cain's and our own self-ownership, lend much comfort to those who deny that parents are at liberty to secure an elective abortion. For the divestiture described above still leaves Cain's parents vested with the ownership of his body until his attainment of agency. Unless Cain's mother's pregnancy is the subject of a surrogacy contract, she is under no correlative duty to refrain from terminating it. And if it is a surrogate pregnancy, her correlative duty is not one owed to Cain, since he has no rights.

On the other hand, it's equally true that closely corresponding things may be said about infanticide and lesser forms of parental violence which might be inflicted on pre-agency Cain. Unless his post-natal nurture is the subject of some kind of fostering contract, his parents lack any correlative duty to refrain from treating him as they wish.[11]

If this is a bullet we must bite as the price of having a non-paradoxical conception of universal self-ownership, it's also true that having it engenders some countervailing considerations. In the first place, a parentally owned pre-agency Cain is better placed than a self-owned one to avoid bearing the liability for any damage or injury which his behaviour may cause to others. Second, Cain's parents' ownership of his body – their lack of any correlative duties to supply it with pre- or post-natal nurturing – does not imply their lack of *non*-correlative duties to do so. Nor, significantly, does it imply the absence of such duties in third parties, within the limits allowed by respect for his parents' property rights. Specifically, those property rights do not preclude more beneficent third parties from pursuing such duties by offering to enter into remunerative surrogacy and/or fostering contracts with Cain's undevoted parents. A third and arguably stronger offsetting consideration is that Cain's ownership by his parents affords him protection against more malevolent third parties. Manipulative persons wishing to remove a credulous Cain from his parents' control, in order to use him in the service of interests which are not his, would be debarred from doing so by the correlative duties they owe to his parents.

So it would appear that, under this dispensation, the fate of pre-agency Cain is utterly dependent on the devotion and resources of his parents and, beyond that, on the distribution of benevolence, malevolence and resources among third parties. But is this true only under *this* dispensation? Could matters be otherwise if pre-agency Cain owned himself?

Surely not. For the various powers needed to exercise his right would, given Cain's non-agency, have to be vested in and exercised by someone else. We might *hope* that any such person would be more benevolent than Cain's parents. But there is no general reason to expect this to be true, and there are overwhelming statistical reasons to believe that precisely the reverse is the case. Hence the casual reassurance which some writers (including Locke) have drawn, from their construal of child custodianship as a relationship of 'trust', can be seen as quite misplaced when we remind ourselves of the extensive discretion placed in the hands of trustees or those who, in legal systems, are empowered to oversee the conduct of trustees. In short, there is no way of freeing the fate of pre-agency Cain – whoever owns him – from dependence on the discretionary decisions of someone else. So if this is a cost of the route by which we arrived at universal self-ownership, it is a cost borne by any other route as well.

Thus far we've been looking at some of the more familiar moral and social issues affected by self-ownership and examining the impact of our derivation of it on them. I want now to venture on to slightly less familiar terrain in order to explore one of that derivation's further and more profound consequences.

The foundation of Cain's (eventual) self-ownership is the use of natural resources in his production. His parents have had to appropriate genetic information from nature to construct his body. In so doing and to retain ownership of him during his pre-agency phase, they must – as do owners of other natural resources – pay compensation to all other persons for the value of what they have appropriated.

But not all genetic endowments are the same. And more to the point, like other natural resources, not all genetic endowments are of equal value. Like other natural resources, their values vary with the values of the uses to which they can be put. The uses to which a natural resource can be put depend upon its physical characteristics and how they interact with other factors of production. The values of those uses depend upon persons' preference orderings. Less valuable or inferior natural resources are ones which can be put only to a set of less valuable uses. Or what comes to the same thing, they are ones which can be put to a set of more valuable uses only by being combined with more valuable (other) factors of production.

The uses to which genetic endowments can be put consist broadly in the production of personal abilities. Such abilities are many and varied, and they include not only things conventionally denoted by action-verbs – skiing, subtracting, singing and the like – but also other capacities normally regarded as aspects of body states, such as an immunity to disease X or an allergy to substance Y. All of these refer to things which a person's body can or cannot do or be.

Of course, whether these abilities are produced will depend – as in any production process – not only on the physical characteristics of persons' genetic endowments but also on the other production factors with which these interact in the course of the body-building process. Factoring out the contribution made to this process by any of these ingredients is a task which we all do impressionistically and which countless researchers in a vast number of fields do professionally. The most that can and needs to be said for our purposes is that a person's genetic endowment constrains the set of abilities that can be produced with it, just as the kind of steel used to build a bridge constrains its load-bearing capacity.

Thus Cain's genetic endowment is inferior to another person's if the cost of using it to produce a given level of ability-value is greater than the cost of using the other person's. The value of an ability is equal to the amount that people would pay for its use or possession, or to the amount of compensation that one person would owe to another for injuring him or her in such a way as to cause the loss of that ability.

Suppose there are only two sets of adults each of which owns one child, Cain and Abel, respectively. And suppose, further, that Cain's genetic endowment is inferior to Abel's. It follows fairly directly that, unless there is an appropriate transfer of wealth from Abel's parents to Cain's, their respective holdings will not be in conformity with the appropriation rule, since that rule prescribes their respective entitlements to an equal portion of natural resource value. Thus, if the value of Cain's genetic endowment is x and that of Abel's is $x + 10$, then what Abel's parents owe to Cain's parents is equal to 5, or in other words, half the difference between the values of their respective genetic endowments. The generalization of this finding is simply that, in a world of many varyingly endowed children, adults owning those with superior endowments must transfer wealth to all other adults.

Does this compensation requirement amount to a welfare fund for (the parents of) disabled children? Not quite. The reason why is that it is utterly unaddressed to the distribution of *ability*. It in no way invites us to pick out those with inferior ability and to compensate their parents by imposing transfer duties on the parents of children with superior ability. Abilities, as was noted, are the joint products of genetic endowments and other factors of production. They are the fruits of production processes which people undertake with factors which they own. Historical Entitlement Theory would offer no greater mandate for the compulsory equalization of holdings in these products, even if they were transferable, than in the case of any other products. Nor, therefore, does it mandate compulsory transfer payments to offset their unequal possession. On the contrary, it expressly forbids them under the principle that persons are not to be deprived of the fruits of their labour. Thus the parents of children with superior ability but inferior genetic endowments owe *no* compensation, while conversely, parents of children with inferior ability but superior genetic endowments *are* subject to a transfer payment duty.

Of course, in a world of fully rectified injustices – a world where private holdings were no longer tainted by past rights-violations including, *inter alia*, inegalitarian detentions of natural resource values[12] – there might be some grounds for regarding the compensation requirement as a welfare fund for (the parents of) disabled children. Why? Essentially because in such a world there is likely to be a much closer correlation between ability-inferiority and genetic inferiority.

The reasoning behind this conjecture runs as follows. In such a world, it would be somewhat surprising if the redistributive effect of such rectification, particularly when secured on a global scale, did not amount to a considerable dispersal of currently concentrated holdings of wealth and a consequent reduction in inequalities of ownership. And it would thus be reasonable to suppose that the effect of this dispersal on the opportunity cost of producing a given level of ability-value in children would be lower for many more persons than those for whom it would increase. Having command over more/better production factors to combine with their children's genetic endowments, more people would have to sacrifice less to produce higher ability-levels than would otherwise be the case. More people would find it easier to bring up children who were healthier, more skilled and better informed. In general, children's ability-differentials would be narrowed. And accordingly, their respective positions at the starting-gate of adult life would be considerably less unequal than they currently are.

To conclude: I began this essay by noting that the right of self-ownership and its derivative entitlement to the fruits of one's labour, though deeply entrenched in our ordinary moral thinking, are also commonly seen as serious obstacles to the redistributive demands of many worthy social objectives. What I hope to have shown is that the universalization of this very right itself mandates a redistribution of wealth that can significantly contribute to the attainment of what is surely among the more important of any such objectives.[13]

NOTES

1 Cf. Robert Nozick, *Anarchy, State and Utopia* (Oxford: Blackwell, 1974), chs 7, 8; G. A. Cohen, 'Self-ownership, world-ownership and equality', in Frank Lucash (ed.), *Justice and Equality, Here and Now* (Ithaca: Cornell University Press, 1986), and 'Self-ownership, world-ownership and equality, part II', *Social Philosophy & Policy* 3 (1986), 77–96.

2 Nozick, op. cit., esp. pp. 169–70, 232–8.

3 Herbert Spencer, *Social Statics*, 1st edition (London: John Chapman, 1851), pp. 133–4.

4 Cf. Sir Robert Filmer, *Patriarcha and Other Writings*, ed. Johann Sommerville (Cambridge: Cambridge University Press, 1991).

5 This suggests the intriguing question of whether 'creationists' are thus logically committed to rejecting, *inter alia*, the American constitution.

6 John Locke, *Two Treatises of Government*, ed. Peter Laslett (Cambridge: Cambridge University Press, 1967), p. 305; Nozick, op. cit., p. 172.

7 Locke, op. cit., pp. 305–6.

8 The first three of these rules are listed by Nozick, op. cit., p. 151. The fourth is not but should be, since it is logically presupposed by much of his argument and is not reducible to or implied by any of the other three.

9 Locke, op. cit., p. 313.

10 As well as the previously mentioned arguments of Locke, Spencer and Nozick, prominent writings on what has sometimes been called 'the land question' include: Thomas Paine, *Agrarian Justice* (1796), reprinted in M. Beer (ed.), *The Pioneers of Land Reform* (London: G. Bell & Sons, 1920); Leon Walras, *Etudes d'économie sociale*, 2nd edition, ed. G. Leduc (Paris: Pichon and Durand-Auzias, 1936), pp. 267–350; and the many works of Henry George, of which *Progress and Poverty* (London: William Reeves, 1884) is the best known.

11 For arguments linking the permissibility (or non-permissibility) of abortion and infanticide, see: Michael Tooley, *Abortion and Infanticide* (Oxford: Oxford University Press, 1983); and L. W. Sumner, *Abortion and Moral Theory* (Princeton: Princeton University Press, 1981) and John Harris, *The Value of Life* (London: Routledge, 1985, 1992).

12 On the compounded injustices entailed by unrectified rights-violations, see my 'Exploitation: a liberal theory amended, defended and extended', in Andrew Reeve (ed.), *Modern Theories of Exploitation* (London and Los Angeles: Sage, 1987), 'Capitalism, justice and equal starts', *Social Philosophy & Policy* 5 (1987), 49–71, and *An Essay on Rights* (Oxford: Blackwell, forthcoming), chs 5, 8.

13 Earlier versions of this paper have been read to seminars at Oxford, Warwick, Bowling Green State and Manchester Universities, at Manchester Polytechnic, at a meeting of the September Group and at a conference on Biotechnology and Ethics (Manchester, 1991). I am greatly indebted to the participants at these gatherings and especially to Jerry Cohen, John Harris, John Roemer, Caroline Steiner and Philippe Van Parijs for their detailed comments.

5

THE MORAL STATUS OF EXTRACORPOREAL EMBRYOS

Pre-born children, property or something else

Bonnie Steinbock

Some of the thorniest problems posed by the new reproductive technologies concern who has jurisdiction over extracorporeal embryos in the event of the parents' death or divorce. The ability to freeze embryos is an important technological advance, as it gives patients several chances at procreation with only one extraction procedure. At the same time, cryopreservation of embryos creates legal problems, and these legal problems raise moral and conceptual questions. What should be done if the couple divorce, and there has been no prior agreement about the disposition of the frozen embryos? When such cases come before the courts, judges will have to decide how to consider extracorporeal embryos. Are the embryos the property of the couple concerned? Or should the embryos be considered to be 'pre-born children' and a custody model employed? This was the issue in the case of *Davis* v. *Davis*.[1]

Mary Sue Davis and Junior Lewis Davis married in 1980. They very much wanted to have a family, but after Mrs Davis suffered five tubal pregnancies, she had her Fallopian tubes severed to prevent further risk to her. She and her husband thereafter decided to resort to *in vitro* fertilization. After six unsuccessful attempts at IVF, the Davises tried to adopt a child. This too was unsuccessful, and they returned to the IVF programme.

In the autumn of 1988, Mrs Davis learned about a new technique, cryopreservation, whereby several eggs could be removed by laparoscopy, fertilized, and then frozen for future use. On 8 December

1988, nine eggs were aspirated from Mrs Davis, fertilised with Mr Davis's sperm, and allowed to mature *in vitro* to the eight-cell cleavage stage. Two of the embryos were implanted in Mrs Davis on 10 December 1988, neither of which resulted in a pregnancy. The remaining seven were placed in cryogenic storage for future implantation purposes.

The Davises discussed the fact that the storage life of the embryos probably would not exceed two years. They also considered the possibility of donating the remaining seven embryos, should Mrs Davis become pregnant as a result of her implant on 10 December, but the couple made no decision about that matter. Nor did they make any document or consent form stipulating how the embryos should be disposed of in the event of future contingencies. (The IVF centre they used had standard consent forms, but as the Davises were 'old customers', having been patients for years, the staff apparently felt it unnecessary to use them.)

The dispute over the fate of the frozen embryos arose when the couple filed for divorce in February 1989. Mr Davis's filing papers requested an order enjoining the fertility clinic from releasing the embryos to Mrs Davis or others for the purposes of thawing and implantation. With divorce impending, Mr Davis did not want to become a parent. Mrs Davis contended that she was the mother of the embryos, and that she had the right to try and establish a pregnancy with them. Moreover, she contended that the embryos were 'pre-born children' with rights of their own.

The *Davis* case thus raises two distinct issues: (1) whose reproductive rights should prevail when, as in the *Davis* case, there is no prior agreement, and one party wishes to avoid reproduction and the other wants the embryos implanted? and (2) how should we think about the embryos? Do they have rights and interests of their own? I will consider the second question first. I will argue that extracorporeal embryos are not the kinds of entities that can have interests or rights. Then I will go on to discuss how we might view such embryos, and what implications this has for resolving dispositional disputes.

I EMBRYOS, INTERESTS AND RIGHTS[2]

In a seminal paper, 'The rights of animals and unborn generations',[3] Joel Feinberg posed the question, what kinds of entities can have rights? That is, what characteristics must a being have for it to make sense for us to ascribe rights to it? Feinberg persuasively argues that not everything can have rights. Stones, stars, ponds – these are not the kinds of things that can have rights. The reason becomes apparent when we consider the function of rights. Rights protect interests. Mere things don't have interests; therefore it doesn't make any sense to ascribe rights to them. The view that interests are necessary for rights-ascription is known as 'the interest principle'.

The next question is, what is it to have interests? Feinberg usefully analogizes having an interest in something to having a stake in it. People generally have a stake in such things as their health, their careers, their assets, their families. In *Harm to Others*, Feinberg develops this idea:

> One's interests, then, taken as a miscellaneous collection, consist of all those things in which one has a stake, whereas one's interest in the singular, one's personal interest or self-interest, consists in the harmonious advancement of all one's interests in the plural. These interests . . . are distinguishable components of a person's well-being: he flourishes or languishes as they flourish or languish. What promotes them is to his advantage or *in his interest*; what thwarts them is to his detriment or *against his interest*.[4]

This way of thinking about interests connects them to what we care about or want, to our concerns and goals, to what is important or matters to us.

If we think of interests as stakes in things, and understand what we have a stake in as defined by our concerns, by what matters to us, then the connection between interests and the capacity for conscious awareness is clear. Without conscious awareness, beings cannot care about anything. They cannot have desires, preferences, hopes, aims or goals. Nothing matters to non-sentient, non-conscious beings. Whether they are preserved or destroyed, cherished or neglected is of no concern to them.

This will be rejected by those who have an animistic conception of the universe, according to which everything is conscious and aware. The stone resents being kicked; the pond is saddened when acid rain destroys its plant life and kills its fish. If this conception of natural objects were right, then the ascription of interests, and therefore rights, to stones and ponds would be perfectly intelligible. But I do not believe that this is in fact the case. I realize that there are cultures, such as certain Native Americans, who disagree, and I can no more prove that they are wrong than I can disprove the existence of an immortal soul. I can only say that an animistic conception of the universe is inconsistent with a scientific view, and the scientific view has a great deal more to recommend it.

Of course, the fact that it does not matter *to* 'mere things' what is done to them does not mean that it does not matter at all. Consider the example of flag-burning. Many people regard this as very wrong. In fact, they get quite incensed about this, and have tried to pass laws protecting the flag. However, proponents of such laws do not think that burning the flag is *cruelty* to flags. They do not think that abuse of a flag is like child abuse, nor would they justify protecting flags with golden-rule arguments ('How would you like to be burned if you were a flag?'). Rather, the flag is viewed as an important symbol of our country, and one that deserves respect. My point is not that there *should* be laws against flag-burning. The answer to that question depends on the nature and value of free speech. My point is rather that such protective legislation does not imply that the entity to be protected has rights.

Unlike flags, embryos are not mere things. They are alive and they are human. For the judge in *Davis* v. *Davis*, W. Dale Young, this was enough to cause him to rule in favour of Mrs Davis. Judge Young framed the issue not as who should get the embryos, but rather whether the embryos were people or products. He concluded that the embryos *in vitro* were people, and therefore that a 'best-interest' analysis was the appropriate one. He held that it was in the manifest best interest of these children that they be available for implantation and that their mother be permitted to bring them to term.

The judge's decision that the embryos are people was based exclusively on the testimony of one witness, French right-to-life physician Jerome Lejeune. The testimony of the other witnesses was rejected primarily because they all termed the embryos 'pre-embryos'. The judge held that he could not find the term in any encyclopaedia or dictionary, and hence concluded that there was no such term. In fact, the term is widely used by biologists because of its greater accuracy in characterizing the initial phase of mammalian and human development. The beginning stages of development after fertilization do not establish the embryo proper, but a feeding layer or trophoblast, which begins to function before the embryonic disc forms. For this reason, the zygote, morula and early blastocyst may be regarded as pre-embryonic stages, with the term 'embryo' reserved for the entity that appears at the end of the second week after fertilization, when the primitive streak, the precursor of the nervous system, appears.[5] It is only after the appearance of the primitive streak that the embryo can be said to be a unique individual, since prior to this time, twinning can still occur. None of this biological information had any impact on Judge Young. On his view, human life begins at conception, so the embryos were human beings, with all the rights of other human beings.

Like so many other right-to-life advocates, Judge Young assumed that the issue was the genetic humanity of the embryos. But no one has ever disputed that the embryos are genetically human. The issue, totally missed by Judge Young, is whether these human embryos have the moral or legal status of born human beings. As law professor John Robertson explains, 'While the preimplantation embryo is clearly human and living, it does not follow that it is also a "human life" or "human being" in the crucial sense of a person with rights or interests.'[6]

To see that being alive and human is not sufficient for the possession of interests, consider the example of an individual in a persistent vegetative state (PVS). PVS patients have suffered severe neurological destruction of the cerebral hemispheres, which contain the function of consciousness or awareness, as well as voluntary action. Because of the damage to their brains, they are permanently unconscious, though not brain-dead. The brain stem of PVS patients remains relatively intact, enabling them to breathe, often unassisted by a respirator, to swallow, and to digest food. They often survive for decades.

Put aside for the moment the possibility of incorrect diagnosis.[7] Assume that the individual really is irreversibly unconscious. I maintain that talk of a right to life here does not make any sense. For what interest is protected by extending indefinitely the biological life of someone who is not and never will again be conscious? No one, I maintain, has an interest in continued existence without conscious awareness of any kind, now or in the future.

On the view I have been defending, moral and legal rights can be intelligibly ascribed only to beings with interests of their own, and this means beings with conscious awareness. The most primitive kind of awareness would seem to be the capacity to experience pain and pleasure, often referred to as sentience. Certainly the ability to feel pain would precede more highly developed cognitive states, such as thoughts, emotions or moods. Precisely when foetuses become sentient is a matter of debate among biologists. Some believe that sentience does not occur until relatively late in gestation, because the neural pathways necessary for pain perception do not develop until the end of the second trimester. Others think that sentience may develop earlier in gestation, perhaps as early as the eighth week of gestation, when brain waves occur. Despite disagreement about precisely when sentience occurs, there is complete agreement that the very early embryo cannot be sentient, because it has not yet developed the rudimentary structures of a nervous system.

It might be conceded that pre-implantation embryos do not *now* have interests. But it might also be said that there is an important and morally relevant difference between embryos and PVS patients, namely that embryos are potential human beings, who will have interests if allowed to develop. The potentiality argument maintains that potential human beings have the same rights as actual human beings, including a right to life.

There are two problems with the argument from potentiality. First, it seems to be a logical mistake to conclude from the fact that something is potentially an x that it has the rights of an actual x.[8] For example, the President of the United States has the right to command the armed forces. Before he took the Oath of Office, George Bush was a potential President of the United States. But he did not then have the right to command the armed forces. By the same token, a potential human being does not have the same right to life as an actual human being. It only potentially has that right, contingent on its becoming a human being.

This logical problem might be overcome by construing the potentiality principle as normative. The claim is not that potential human beings now have a right to life in virtue of their potentiality, but rather that their potential to become people provides us with reasons for valuing and protecting their lives. This seems plausible. If one values daffodils, then one ought to protect daffodil bulbs, even though bulbs are not daffodils. However, another problem remains. Why shouldn't sperm and ova also be considered potential human beings, with a right to life? Philosopher John Harris makes the point this way:

> To say that a fertilized egg is potentially a human being is just to say that if certain things happen to it (like implantation), and certain other things do not (like spontaneous abortion), it will eventually become a human being. But the same is also

> true of the unfertilized egg and the sperm. If certain things happen to the egg (like meeting a sperm) and certain things happen to the sperm (like meeting an egg) and thereafter certain other things do not (like meeting a contraceptive), then they will eventually become a new human being.[9]

So, sperm and ova also have the right to life, and the use of Delfen foam is mass murder. Few defenders of the potentiality principle are willing to accept this, and so they try to show that there are reasons why a fertilized egg is a potential human being but sperm and ova are not.

What reasons? Potentiality theorists often point out that fertilization marks the beginning of an ongoing process which, if it is not deliberately interrupted, has a pretty good chance of resulting in the birth of a baby. By contrast, they note, a gamete is not developing into anything. Unless someone intervenes and causes the gamete to conjoin with another gamete, it will simply pass out of the body and die. Whatever appeal this argument has in the context of normal fertilization and pregnancy, it does not apply in the case of extracorporeal fertilization. The fertilized egg *in vitro* cannot develop into a foetus all by itself. Unless someone intervenes, and transfers the embryo into a uterus, it too will die. If it is claimed that gametes are not potential people, because they will not, on their own, develop into human beings, then it must be acknowledged that precisely the same is true of extracorporeal embryos.[10] At least in the context of IVF, there does not seem to be an enormous moral difference between the zygote in the petri dish just after fertilization has occurred and the sperm and the ovum in the petri dish just prior to fertilization.

To show how ludicrous it is to equate embryos with children, bioethicist George Annas gives the following example.[11] If a fire broke out in a laboratory where the seven embryos were stored, and a two-month-old child was also in the laboratory, and only the embryos or the child could be saved, would anyone hesitate before saving the child? Of course not. This shows that no one really does equate embryos and children, and the absurdity of a 'best-interests' analysis applied to blastocysts.

Annas notes that if the judge really believed that he had to decide this case based on the 'best interests of the children', he would have had at least to determine if Mrs Davis was a fit mother to gestate them.

> Given her past history of inability to carry a fetus to term, there is little probability of her successfully gestating any of the seven embryos. Requiring her to hire a surrogate mother to gestate them would almost certainly enhance their chances to be born.[12]

Annas also points out that despite the fact that Judge Young spent all his time deciding that the embryos were people, not property, he ended up treating them like property. 'Instead of deciding custody, visitation, and support issues (which he would have to do if the embryos *were* children), he awards them to Mrs Davis in exactly the way he would award a dresser or a painting.[13]

Judge Young's decision was bad law and bad bioethics. This was recognized by the Tennessee Court of Appeals, an intermediate-level appeals court, which overturned Judge Young's decision. The ruling was widely reported as giving both Junior Davis and his ex-wife, who has since remarried and is now Mary Sue Stowe, 'joint custody' of the embryos. In fact, Mrs Stowe and Mr Davis were given 'joint *control* of the fertilized ova (and) equal voice over their disposition'.[14] The distinction is important. Joint custody suggests that the embryos were children; joint control does not. The appellate ruling was upheld in a unanimous decision by the Tennessee Supreme Court on 1 June 1992.[15]

If the embryos are not 'pre-born children', as Mrs Stowe maintained, are they property? If so, can they be sought and sold, as sperm can be sold? Can they be marketed for use in cosmetics?[16] Many people, not just right-to-lifers, find such ideas to be extremely repugnant. Fortunately, there is a third alternative to viewing extracorporeal embryos either as children or as property. This third position was taken by the Tennessee Supreme Court, which said, 'We conclude that the preembryos are not, strictly speaking, either "persons" or "property" but occupy an interim category that entitles them to special respect because of their potential for human life.' A similar view was taken by the Ethics Advisory Board when it concluded that 'the human embryo is entitled to profound respect; but this respect does not necessarily encompass the full legal and moral rights attributed to persons'.[17] John Robertson, who wrote the brief for Mr Davis, has taken this position, maintaining that the human embryo can be accorded respect, not for what it is in its own right, but for the symbolic meaning it has for us:

> Although neither a person nor an entity possessing interests, it

> may be the subject of duties created to demonstrate a commitment to human life and persons generally. Justice may not require that we grant the embryo rights, but we may choose to treat the embryo differently than other human tissue as a sign of respect for human life generally.[18]

George Annas is another bioethicist who takes this third position, maintaining that Judge Young's person/product dichotomy was very misleading:

> embryos could just as easily be considered *neither* products nor people, but put in some other category altogether. There are many things, such as dogs, dolphins, and redwoods that are neither products nor people. We nonetheless legally protect these entities by limiting what their owners or custodians can do with them. Every national commission worldwide that has examined the status of the human embryo to date has placed it in this third category: neither people nor products, but nonetheless entities of unique symbolic value that deserve society's respect and protection.[19]

This respect and protection can be shown in two ways. First, by limiting the time in which extracorporeal pre-embryos can be used, whether in research or for conception. Most commissions have chosen fourteen days after fertilization as a cut-off, since that is when the primitive streak appears. Second, by insisting that the uses to which pre-embryos are put are medically significant, and not frivolous.

Although Annas and Robertson agree that embryos are neither people nor products, there is a subtle dispute between them. Where Robertson emphasizes the optional nature of respect for embryos, saying that we may *choose* to treat the embryo differently as a sign of respect for human life, Annas maintains that extracorporeal embryos *deserve* society's respect and protection. Presumably Annas does not think that embryos have a *right* to social protection, but rather he would regard a society which did not choose to protect embryos as morally deficient.

The difference between Annas and Robertson becomes less important once we understand that they are approaching the issue from quite different perspectives. Robertson appears to be taking an external or sociological perspective. He sees that most people in our culture regard embryos differently from individual gametes. Given that embryos lack interests and cannot be harmed, this attitude towards embryos is not rationally required, but nevertheless since people feel that way, their sensibilities should be respected. Annas, on the other hand, is taking an internal view. He is trying to explain and justify the conviction of many people in our society that extracorporeal embryos have a special moral worth. Such an explanation appeals to the fact that although embryos are not yet people, they are not just like gametes either. Because fertilization ordinarily marks the beginning of a process that can result in the birth of a new human being, the pre-implantation embryo is seen as a symbol of human life. It has a symbolic value that no gamete by itself has, and for this reason ought to be protected by a society that values human life.

II RESOLVING DISPOSITIONAL DISPUTES

What are the implications of this third alternative for deciding *Davis* v. *Davis*? The interests to be considered are not those of the pre-implantation embryos, for pre-conscious, pre-sentient entities do not have interests of their own. Rather, the relevant interests belong to the couple. The question is whose interests are most important and should prevail: Mrs Stowe's interest in becoming a mother or Mr Davis's interest in avoiding becoming a father?

While acknowledging that both parties have significant interests, John Robertson argues that, in absence of advance instructions, the party wishing to avoid reproduction – in this case, Mr Davis – should prevail. Robertson argues that Mr Davis would be irreversibly harmed if embryo transfer and birth were to occur, as he would be forced to accept the psycho-social and financial burdens of parenthood.

The risk of financial liability could be removed by statute, relieving the party wishing to avoid reproduction of rearing rights and responsibilities, as has been done with donor insemination. As things stand now, however, a man providing sperm for insemination in the IVF context is the legal father, with rearing rights and duties, including support requirements. Mrs Stowe has said that she has no interest in securing child support from her husband, but that would not prevent the children themselves or the state from seeking child support at a future date.

Mr Davis's primary objection to becoming a father was not fear of financial liability, but rather what Robertson refers to as the 'psychosocial impact of unwanted biologic offspring'. Mr Davis objected to being 'raped of his reproductive rights' and also to having a child produced to live in a single-parent home. His own life was shattered at the age of 6 when his parents were divorced, and he and his three brothers were sent to a boys' home. These psychosocial burdens are unavoidable if Mr Davis becomes a father against his will. By contrast, Robertson says, Mrs Stowe is not irreversibly harmed by being denied embryo transfer, as she can reproduce at a later time with other embryos. Admittedly, she has undergone many painful, physically tiring and emotionally taxing procedures. This is not determinative, according to Robertson, 'since the burdens of any one additional retrieval cycle are moderate and acceptable, at least relative to the irreversible burdens of imposing fatherhood on the husband'.[20]

The question of whose interests should prevail is extremely difficult to resolve when the procreative liberty of both parties is at stake. If these frozen embryos in fact represent Mrs Stowe's last chance to give birth, it would seem that her desire to become a mother should be given as much weight as Mr Davis's desire to avoid fatherhood. The case would have to be settled by attempting to determine which party would be more badly harmed by frustration of his or her reproductive interests.

It is not at all clear to me that Robertson is right to maintain that such cases should ordinarily be decided in favour of the party wishing to avoid reproduction. It seems rather cavalier to say that Mrs Stowe can always reproduce at another time with other embryos. This ignores the reality of *in vitro* fertilization, which involves serious physical burdens. These include being subjected to drugs and hormones to induce superovulation, the long-term effects of which have not been determined; laparoscopy, which carries a significant risk of mortality or morbidity; and the risk of infection, laceration of the uterus or an ectopic pregnancy.

In addition to the physical burdens, IVF imposes personal and psychological burdens, including financial loss, loss of privacy, and the emotional trauma when a pregnancy does not occur or is spontaneously aborted. Mrs Stowe's willingness to bear these burdens indicates the intensity of her desire to have a child, and the pain she would experience if irrevocably prevented from fulfilling her desire. Moreover, the desire to have children is one with which most people can sympathize and understand.[21] It is arguably both more intense and more important than Mr Davis's desire that there should not exist genetic children of his to whom he will have, by his own choice, no rearing ties.

However, the situation changed when Mrs Stowe decided not to try to have the pre-embryos implanted in her uterus. Her lawyer says that she now wants to retain custody of the pre-embryos because they are 'potential life' and she wants that potential realized. An appellee brief filed by her lawyer in May 1990 said she wants to donate the eggs to another infertile couple. This move prompted one of the judges on the state appeals court panel to question Mrs Stowe's motives for still wanting the embryos. 'Is this a case of a party wanting to win at any cost?' Judge Franks asked.[22]

From a right-to-life perspective, Mrs Stowe's motives are noble. Her concern is solely for the welfare of her 'pre-born children'. She is willing to renounce her claim to the frozen embryos, and her chance to become a mother, in order to enhance their chance of live birth. (The chance of a live birth resulting from these embryos, which have been frozen for almost four years, is remote.)

The interest principle requires a completely different approach. It maintains that the welfare of the embryos is not the issue, because fertilized eggs do not have a welfare or interests of their own. The only interests are those of the adult parties. The most important aspect of the *Davis* case is that it no longer involves a conflict of interest in reproductive liberty. There is only one party whose reproductive liberty is at stake, and that is Mr Davis, who wishes to avoid paternity. Mrs Stowe has no reproductive interest, because she no longer intends to try to reproduce with the embryos. Her interest in having her genetic offspring brought to birth by someone else, who will then become the rearing parent, should have no weight at all. The decision of the Tennessee Supreme Court was correct in upholding Mr Davis's right to procreational autonomy. On 10 June 1993 Junior Lewis Davis disposed of the embryos.

Obviously, it would be better for everyone concerned if such matters never, or rarely ever, reached the courts. Instead, IVF programmes should require that couples are adequately counselled, and that they create advance directives, if they are unavailable or unable to agree on disposition when death, divorce or other contingencies occur. This will not entirely resolve the problem, since advance directives, like wills, can be challenged. Still, requiring people to think about future contingencies and to make decisions in advance will prevent most cases from ending up in court. When they do, judges should not use such cases as a forum for expressing anti-abortion sentiments. They should recognize that the relevant interests are those of the disputing parties. These cases will rarely be easy to resolve, but without conceptual clarity regarding the nature and status of extracorporreal embryos, they will be hopeless.

NOTES

1 Junior L. Davis vs. Mary Sue Davis vs. Ray King, M. D., d/b/a Fertility Center of East Tennessee, *In the Circuit Court for Blount County, Tennessee, at Maryville, Equity Division (Division I)*, No. E–14496, 21 September 1989.

2 Some of the material in this section comes from the first chapter of my book, *Life Before Birth: The Moral and Legal Status of Embryos and Fetuses* (New York: Oxford University Press, 1992), where I develop a theory of moral status that I call 'the interest view' based on Joel Feinberg's 'interest principle'.

3 Joel Feinberg, 'The rights of animals and unborn generations', in William T. Blackstone (ed.), *Philosophy & Environmental Crisis* (Athens, Georgia: University of Georgia Press, 1974), pp. 43–68. Reprinted in Thomas A. Mappes and Jane S. Zembaty (eds), *Social Ethics*, 3rd edition (New York: McGraw-Hill, 1987), pp.484–93.

4 Joel Feinberg, *Harm to Others* (New York: Oxford University Press, 1984), p.34.

5 John A. Robertson, 'Embryos, families, and procreative liberty: the legal structure of the new reproduction' (hereafter, 'The new reproduction'), 59 *Southern California Law Review* 939 (1986), 969–70.

6 John A. Robertson, 'Resolving disputes over frozen embryos', *Hastings Center Report* 19:6 (November/December 1989), 11.

7 There have been a few cases in which the diagnosis of PVS was incorrect. See my 'Recovery from persistent vegetative state? The case of Carrie Coons', *Hastings Center Report* 19:4 (July/August 1989), 14–15. Despite such cases, most neurologists regard diagnoses of persistent vegetative state as among the most reliable in medicine.

8 The claim that the potentiality principle involves a logical mistake was first made by Stanley Benn in 'Abortion, infanticide, and respect for persons', in Joel Feinberg (ed.), *The Problem of Abortion*, 2nd edition (Belmont, Calif.: Wadsworth Publishing Company, 1984), p. 143.

9 John Harris, *The Value of Life: An Introduction to Medical Ethics* (London: Routledge & Kegan Paul, 1985), pp. 11–12.

10 Peter Singer and Karen Dawson, 'IVF technology and the argument from potential', *Philosophy & Public Affairs* 17:2 (spring 1988), 87–104.

11 George J. Annas, 'A French homunculus in a Tennessee Court', *Hastings Center Report* 19:6 (November/December 1989), 22.

12 ibid.

13 ibid., 21.

14 Duncan Mansfield, 'Joint custody granted in embryo appeal', Associated Press, *The Times Union*, Friday 14 September 1990, A–7.

15 Ronald Smothers, 'Court gives ex-husband rights on use of embryos', *New York Times*, 2 June 1992, A1.

16 'Embryos to lipsticks?', *New Scientist*, 10 October 1985, 21. Cited in Alan Fine, 'The ethics of fetal tissue transplants', *Hastings Center Report* 18:3 (June/July 1988), 7.

17 Ethics Advisory Board, Department of Health, Education and Welfare, *Report and Conclusions: HEW Support of Research Involving Human In Vitro Fertilisation and Embryo Transfer* (Washington, D.C.: US Government Printing Office, 4 May 1979), pp. 35–6.

18 Robertson, 'The new reproduction' 974.

19 Annas, op. cit., 20. Annas lumps together dogs, dolphins and redwoods. By contrast, the interest view differentiates dogs and dolphins from redwoods, on the ground that while none of them are persons, dogs and dolphins nevertheless can have interests and a welfare of their own. They are, for this reason, eligible for rights-possession. The capacity to have interests differentiates animals from flags, redwoods and embryos. The grounds for protecting sentient and non-sentient beings are importantly different, on the interest view.

20 Robertson, 'Resolving disputes over frozen embryos', 9.

21 Some feminists argue that the desire to become pregnant is neither genuine nor free, but is conditioned by society's pro-natalist ideology and the patriarchal power structure with which it is associated. See, for example, Gena Corea, *The Mother Machine: Reproductive Technologies from Artificial Insemination to Artificial Wombs* (New York: Harper & Row, 1985). For an opposing view, see Mary Anne Warren, 'IVF and women's interests: an analysis of feminist concerns', *Bioethics* 2:1 (January 1988), 37–57.

22 'Chill in custody fight', *National Law Journal*, 18 June 1990, 6.

6

IVF AND MANIPULATING THE HUMAN EMBRYO

Susan Kimber

INTRODUCTION

One might assume that all right-minded people agree that pursuing a 'better' quality of life is a worthwhile goal. The problem is they will not necessarily agree about what makes for such quality, what means can be justified in approaching it, or where the balance is to be found between the improvement of life for the individual and that for society as a whole. In this respect, the field of reproduction and fertility, like other areas of medicine, is caught in tension between those advocating the newest advances medical research can offer to promote reproductive health for all and those who counsel caution in the implementation of new and relatively untested techniques.

Since the setting up in 1982 of the Committee of Inquiry into Human Fertilization and Embryology under Dame Mary Warnock (the Warnock Committee), to consider the then new developments in infertility treatment and their repercussions, the topic of *in vitro* fertilization (IVF) has seldom been far from public debate. The parallel developments in molecular biology allowing analysis and modification of the genetic material, DNA, have, in the opinion of some, opened a Pandora's box riddled with ethical pitfalls surpassing even the technical ones.

The marriage of these two technical advances presents us not only with the possibility of carrying out straightforward fertilization of an egg outside the body, but also with the possibility of deliberately changing the genetic make-up of gametes or embryos.

To cover all the possible areas of ethical concern surrounding IVF and related reproductive technology would require more than a single essay; therefore it has been necessary to restrict this chapter to certain key issues, leaving many holes in the debate and presenting only a cursory treatment of a number of questions. This is inevitably a personal view of what can and will be possible in this area of reproductive medicine. I will discuss some of the psychological and ethical problems confronted by both patient and medical or scientific practitioners and indicate my view of the limits of what is permissible and feasible in this area, both now and in the future.

IVF AS A CURRENT MEDICAL TREATMENT

The technique of IVF is now so well known that it hardly needs to be described. The woman is generally given hormone treatment (mimicking the changes in natural hormones controlling the ovary) to stimulate growth of several eggs (oocytes) in the ovary. In the natural course of events usually only one egg grows to maturity and is ovulated at each cycle. A final injection stimulates ovulation, and the eggs are generally collected surgically. The release of several eggs increases the chance of successful fertilization. After a period of maturation in a dish containing a defined culture medium, sperm are added and fertilization takes place. The fertilized eggs may be allowed to undergo one or more cell divisions in culture before being replaced in the woman's uterus. I shall refer to pre-implantation stages following the union of sperm and egg as embryos throughout this chapter. The term 'pre-embryo' has been introduced for early stages of development between fertilization and implantation, but it has not been adopted widely in the scientific community. It has no real biological meaning. The term has been used mainly to ease the consciences of those who are uneasy about carrying out research post-fertilization and to pacify the lay-public by employing a less emotive term, without (in my opinion) facing, or helping the general public to face, the real issues. I do not see this term becoming incorporated permanently into the scientific literature and will not use it here.

IVF is one treatment for infertility and must be seen in the context of reproductive medicine as a whole. I know of no gynaecological clinic with an IVF facility where the use of fertilization outside the body (involving as it does invasion of the woman's body for the collection of eggs, often surgically and under anaesthetic) is anything more than *one of a number* of treatments available for problems in conception. It is by no means ever the first line of therapy, and in many cases would be a totally inappropriate means of alleviating the problem preventing conception.

Medical states which IVF *may* effectively alleviate fall into the following categories:

1 Tubal blockage, which is the predominant problem for which IVF has been used as the solution, and accounts for 70 per cent of cases.
2 Immunological problems: for instance the woman may have antibodies to her partner's sperm.
3 Oligospermia (few sperm).
4 Idiopathic infertility (cause unknown).
5 Anovulatory cycles.

Categories 2–5 make up most of the remaining 30 per cent of IVF cases at present. For some of the couples facing these problems gamete intra-fallopian transfer (GIFT), in which the gametes are placed in the Fallopian tubes, may be an appropriate alternative.

In the future, couples who experience none of these problems but who have been identified as carriers of severely debilitating or life-threatening genetic diseases will no doubt also fall into the category of those able to be helped by IVF (see below). There is also, or so the press would have us believe, the possibility of 'convenience IVF'. A couple may have eggs or embryos frozen when the woman is in her twenties, in order to initiate a pregnancy at a later date when it would not be so disruptive to the furtherance of the woman's career. This would avoid the problem of the increased proportion of chromosomal abnormalities found in foetuses carried by women conceiving after their mid- to late thirties.

However, IVF is not something to be undertaken lightly or, I would contend, for purposes of convenience: the treatment is laborious, time-consuming and expensive and can be severely uncomfortable if not painful for the woman. Furthermore there is evidence to suggest that the ability of a woman's physiological mechanisms to support a pregnancy also decreases with age, particularly from the mid-thirties (Edwards 1980; Cohen-Overbeck *et al.* 1990).

Even more significantly, the overall success rate from IVF (where patients have *not* been carefully selected to improve the statistics!) is still only about 10–20 per cent of embryos replaced leading to live births. This does not take into account the cases where no viable embryos are obtained because of problems on either the male or female side or in the fertilization process. Nor does it take into account the fact that most clinics have periods during which the success rate may drop to almost zero, often for no apparent reason. Furthermore the success rate from frozen embryos may be even lower.

It has often been questioned whether infertility is indeed an illness, and whether the considerable resources needed to carry out IVF should be directed towards what is certainly neither a life-threatening disease nor even, superficially, a minor ailment. Women and men attending IVF clinics are in general healthy: probably in better health than average members of the world population. But this argument must surely be flawed in the light of the effect of infertility on the marriage relationships and mental health of at least some individuals. Absence of children has been suggested as the cause of nervous breakdown, marriage break-up, and even suicide, as well as exacerbating other physical and psychological problems in some but not, of course, all those affected. One might just as well apply the resources argument to other situations leading to break-down of mental health. Most psychiatrists would surely agree that it is better to remove the source of a psychological problem than to maintain the patient on valium for the rest of her or his life.

There are *inevitably* limited resources, and we live in a world where millions of people still lack a basic standard of hygiene or immunization against severely debilitating but preventable diseases. However, if we seek to promote the best physical, mental and social health for all human kind, then IVF should surely be a competitor for these funds along with the rest. What proportion of the money available should be put into this type of treatment is another question altogether.

IVF has acquired its highly controversial profile predominantly because of the following characteristics:

1 Its novelty.
2 Its elitism.
3 The fear that it would lead to a high incidence of abnormal embryos.
4 The separation of what the Church of England Board for Social Responsibility (1985) have called (with reference to the Book of Common Prayer) the procreational and the relational goods of marriage, i.e. the creation of children divorced from sexual intercourse.

5 The almost universal generation of extra or so-called 'spare' embryos *in vitro* which are deliberately *not* replaced into the uterus of the woman in order to maximize the probability of a successful pregnancy and reduce the risk of a multiple birth.
6 The possibility of manipulation of the embryo, both by the biopsy of cells for diagnostic purposes and in other ways such as by the alteration of the genetic material.

I will assess these problems briefly.

1 Its novelty

The introduction of a concept so unfamiliar to the general public bred suspicion and fear of exploitation by the unscrupulous. Some of them imagined 'Brave New World' scenarios.

This problem may be countered by remembering that all new techniques have probably been viewed in this way at their inception. The response to the unknown and uncertain is frequently fear, but just because there is the potential for the use of a technique for evil purposes does not mean that its use for what society considers beneficial and positive purposes should be precluded. We do not forbid the use of electricity because of the risk that some mad dictator may use it for the purpose of torture by electric shock.

If we value progress in medicine, then we have to risk teething problems and failures during the initial practice of new treatments whilst trying to minimize them. Transplant surgery is frequently quoted as an example here. Of course the risk must be weighed against the benefit and found acceptable. Society guards against the exploitation of new developments by introducing safeguards such as legislation and the vetting and licensing of practitioners. Such safeguards were recommended in the Warnock Report in 1984[1] and have finally been brought into force with the publication of the Human Embryology and Fertilization Act 1990.[2] Of course the problem is that it may never be possible to devise a range of safeguards which are satisfactory to more than a minority of people.

After publication of the Warnock Report a voluntary licensing authority was established. This comprised a group of scientists, clinicians and lay people who undertook to regulate this area of medicine and research in the light of public concern. The body acted to oversee all proposals for work associated with IVF on a purely voluntary basis and was superseded by an interim Licensing

Authority. In August 1991 the statutory Human Fertilization and Embryology Authority took over officially to supervise the enforcement of the Act, becoming responsible for the granting of licences to clinics and hospitals in all areas of research and clinical practice relating to IVF.

2 Its elitism

IVF is still far from available to all at present. This is first, because IVF clinics are not present in all hospitals and second, because the high cost of the treatment precludes its application on demand, even if it becomes established as a government-funded treatment. If society considers that health care should be available to everyone without discrimination and irrespective of means, this must be a far from ideal situation. With the rapid proliferation of clinics and wider availability of the technology, access must improve and costs must come down.

3 Abnormalities

The fear that IVF will result in the birth of abnormal babies has proved to be unfounded. The incidence of abnormalities among the thousands of babies born following IVF has been slightly lower than that from non-IVF births. The level of abnormality we are considering here is, for instance, 0.02 per cent for embryos with a single X-chromosome (Turner's Syndrome) or 0.15 per cent for trisomy of chromosome 21 (Down's Syndrome), the commonest cause of mental retardation. Considering the generally higher average age of the women involved and the increase in chromosomal defects with age, this suggests a positive advantage of IVF. To the embryologist the lack of abnormalities is not surprising because it is established in other mammals that damage to the embryo before implantation does not have a teratogenic effect (i.e. lead to malformed organs or tissues): it either kills the embryo or can be compensated for, leading to perfectly normal development.

4 The creation of children divorced from sexual intercourse

In the light of contraception and planning in the timing and number of children, as well as the routine loss of at least 60 or 70 per cent of fertilized embryos in the first three months of 'natural' pregnancies (mainly in the first month before pregnancy is even recognized), this separation is already apparent to most couples. It does not present any additional problem unique to IVF.

In the Church of England, as in other denominations of the Christian church, members are far from agreed in their attitude to IVF. The disagreement is even greater when members consider some of the consequences and repercussions of the technique. However, the Board concluded in its Working Party report on Human Fertilization and Embryology (1985) that attempts to relieve childlessness by means of new technological methods assisted marriage to fulfil one of its natural ends as laid down in the Book of Common Prayer. The essence of the arguments put forward revolve around the family as a God-given institution for the fulfilment, mutual assistance and co-operation of a monogamous couple and the ideal setting for the rearing and development of children.

Similarly most other religions have an extremely positive attitude to the family. The predominant view of the majority of people without any kind of religious belief seems to be that the nuclear family is a generally conducive environment for children. This is in spite of the fact that things frequently go wrong and both the divorce rate and the percentage of single parents continue to rise. A first failure does not seem to prevent frequent remarriage among divorcees, with re-establishment of two-parent families, while the all-too-observable struggles of single parents seem to militate against their state being anything but second best.

It can be argued (as pointed out by Dr Pauline Webb recently[3]) that for a child to grow up in an environment where it can learn from, and be cared for by, both a like and a non-like parent in terms of gender will help to equip that child for the real world in which it will have to deal with, and relate to, both men and women. I conclude, as many have done, that to grow up in a stable family with both biological parents is the ideal situation.

Couples resorting to IVF in their desire for a child can suffer mental anguish, physical pain, inconvenience and often considerable loss of dignity on the part of the woman. They invest a considerable amount of time and money, sometimes over many years. This may be a prelude to perhaps greater love and commitment to the IVF-derived offspring than is present in many homes with children conceived by natural means. The IVF child is in this context a product of a loving relationship and the depth of that love is revealed in the lengths that parents are prepared to go to achieve the missing element in their family. We may conclude that IVF is to be welcomed as an additional form of reproductive therapy, an extension to other well-established types of assistance with the reproductive process such as Caesarean section and hormonal induction of labour, to name but two.

But the further and inevitable consequences of the use of IVF and related techniques (discussed below) must also be dealt with. We cannot simply say 'Yes, IVF is a good thing' and hope that the practical repercussions will go away: most of them cannot be avoided.

I have up to now considered IVF or related techniques within marriage. If it is held that the nuclear family is the ideal environment for the rearing of children, then there should be no difficulty in the application of IVF to married couples consenting to such treatment. The child will be the natural genetic offspring of both parents and brought up by both. But what happens if one parent cannot provide gametes? Should an anonymous donor be able to father a child by AID (artificial insemination by donor)? This technique may, of course, be used with or (as has been the case for many years) without IVF, depending on the reproductive health of the woman. Does the introduction of a third party into the relationship in this way constitute adultery? Not if it is considered that adultery involves a breach of faith leading to the sexual union between one member of a couple and another individual not her or his spouse. Of course, some will consider that adultery in this context is of no relevance in any case. But a third party will always be present as an additional shadowy background figure in the husband–wife relationship and in the parent–child relationship. How will the husband cope with the knowledge that he is not the biological father? Will the child be told about her or his origins, and if so, how will he or she cope with the knowledge that 'Dad' is not actually Dad at all . . . or is he? After all, the important nurture element of the parental relationship may well have been fulfilled more caringly and apparently successfully than in many families where fertility intervention has not been necessary. Similar questions need to be asked with respect to donation of oocytes.

Lack of knowledge of true 'roots' and genetic origins *need not* be an insurmountable barrier: it is often an element in adoption, a process which we do not condemn. However, adopted children frequently have problems stemming from their family status and often express a desire to seek out their biological parents. These potential problems may require counselling of parents and offspring and definitely need sensitive handling. The balance between the 'for' and 'against' of AID will rest on the maturity and personalities of the parents, and the careful handling of the additionally complicated family relationships. The child conceived by AID is still a creation of the marriage and will still be directly genetically related to the mother who has carried it in pregnancy and gone through the bonding process with the child in both pre- and post-natal life. As such it might be seen that AID is in fact preferable to adoption, though of course the adoption of already existing children into a loving and caring home cannot be anything but good. Other real concerns are the number of children derived from the sperm of any one donor and the risk of unwitting incest between half-brothers and half-sisters. The former problem has been safeguarded in current legislation by restricting the use of the sperm from each donor, and the probability of the latter is reduced to what most would consider a minor risk by this legislation.

The issue of surrogacy and AID on demand for single women has also attracted considerable public interest and concern over the last few years. Indeed the latter has recently come to the fore with the publicity given to two single heterosexual women who have undertaken AID in clinics in the Midlands. In the light of the arguments already outlined I cannot see that pregnancy by AID for a single woman is a propitious pathway for the child (or even the woman, although she at least has the ability to make a conscious choice). However, whether it should be *legislated* against or *counselled* against is another matter. Most Christians believe that deliberately seeking single parenthood goes against the will of God for humankind,[4] and as such is far from the ideal state for adult or child. It can be argued from observation of the problems of children from single-parent families that this is borne out among real people in real situations.

Surrogacy is another issue stemming from IVF which has given serious cause for concern.[5] Among the practical problems are that of bonding *in utero* between the foetus and the 'carrier' mother, with subsequent confusion for both infant and surrogate mother as well as anguish on the part of the biological mother. The difficulty that the surrogate mother will face in giving up the child or her possible refusal to do so has to be faced. The issue of exploitation is not really tackled by legislating against the setting up of profit-making agencies 'marketing' surrogate mothers. I am sure that individuals can exploit as efficiently as agencies, especially but *not exclusively* when money changes hands. That strong-minded selfish people with burning desires for particular personal goals will exploit weaker, needy or self-sacrificing individuals is unfortunately a characteristic of the human condition. A just society which protects the weak provides safeguards against such exploitation, but the danger will then be that individual freedom will be limited. I do not pretend to be wise or knowledgeable enough to suggest the position of the boundaries between what is inadvisable but permissible and what should be legislated against.

5 The almost universal generation of extra or so-called 'spare' embryos *in vitro*

In most clinics, when fertilization and normal cleavage of several eggs has occurred, a maximum of three embryos are replaced in the uterus (Van Smeirteghem and Van den Abbeel 1990) to maximize the probability of a successful pregnancy whilst reducing the risk of a multiple birth. Any remaining embryos are not replaced.

Should these embryos be frozen for possibly unnecessary future replacement, as insurance against the current pregnancy failing or because (or in case) more children are desired later? Should they be frozen and 'banked' for donation to couples where pregnancy even mediated by IVF is impossible or where one or both partners are carrying a severely deleterious genetic defect? Can one justify using them for research, and if so, are there restrictions on the kind of research that it is desirable to undertake (Edwards 1982a and b)? Should spare embryos be 'allowed to die', or is this more morally objectionable than their incorporation into a valuable research programme that will lead to improvements in IVF or alleviate suffering in other ways?

The Human Embryology and Fertilization Act 1990 has laid down strict legislation on these matters. Before considering the issue of research, I shall deal with the arguments for preserving 'spare' embryos by freezing.

Freezing

The 1990 Act[6] specifies that human embryos may be stored (and then only with appropriate consent) for not longer than five years and gametes for not longer than ten years.

Embryos are frozen in about half the clinics undertaking IVF in this country. However, results from a recent world survey (Van Smeirteghem and Van den Abbeel 1990) revealed that in those centres that responded (mainly the bigger and better-established ones), 75 per cent of supernumary embryos were frozen. Thus freezing is already well established as an adjunct to IVF. In some countries such as Germany and Denmark, the freezing of embryos is forbidden, although the freezing of oocytes is still allowed in Germany. Whether this is an ethically correct decision is doubtful, since there is evidence that the special disposition of the chromosomes in the growing and maturing oocyte may cause them to be particularly susceptible to damage from the freezing procedure or the use of cryoprotectants (substances added to the freezing solution to prevent ice-crystal damage) (Magistrini and Szollosi 1980; Johnson and Pickering 1987; Pickering and Johnson 1987). However, in experiments on mouse oocytes it has not so far been proven that there is any increase in aneuploidy (abnormal chromosome numbers) after fertilization of frozen oocytes. All the same, the success of fertilization of oocytes after thawing is extremely low, and by the end of the 1980s no more than four babies had been born from frozen, thawed and subsequently fertilized oocytes (Chen 1986; Van Uein *et al.* 1987).

The freezing technique is now quite successful for human embryos, and those which have been thawed and replaced have approximately a 2–5 per cent chance of being delivered as healthy babies (Van Smeirteghem and Van den Abbeel 1990; Ahuja 1990). This may be marginally less than the success of immediately transferred embryos, but it depends on a number of factors, including the experience of the practitioner.

Overall the chances of a woman getting pregnant after IVF are increased if she has embryos frozen, because the chances of getting pregnant at each transfer are much the same. If the woman does not have to go through further gruelling rounds of hormonal stimulation and laparoscopy prior to transfer, she is certainly spared much. However, the opinion of some clinicians is that the 'second chance' philosophy is less useful in practice than it might seem: if good embryos are produced and the woman's physiology is receptive to implantation of the embryo, she will become pregnant at the first replacement. Some consider that if she does not become pregnant then she will, in all probability, not do so with a second replacement using frozen and thawed embryos from the same batch. Published evidence for this is lacking, and there is disagreement between clinics.

In the case of a pregnancy from the original replacement the frozen embryos may well become an embarrassment. What is to be done with potentially unwanted frozen embryos is a particularly difficult issue to resolve satisfactorily. At present, in the UK, they are likely to be retained for a period agreed with the couple concerned (up to five years under the 1990 Act) and then destroyed. But it is clear from the survey referred to above (Van Smeirteghem and Van den Abbeel 1990) that about 5 per cent of IVF embryos world-wide are used for donation, presumably with 'parental' consent, and no doubt some frozen embryos will also be thawed for this purpose.

In instances where further children are desired the freezing technique certainly avoids a woman having to re-experience much of the most unpleasant side of IVF. However, at the point of freezing the decision of whether the family is to be extended should the initial pregnancy be successful may not be clear: the birth of twins or triplets may change the mind of even the most decided!

Most people will be familiar with the legal and ethical dilemmas surrounding the scenario of both parents of a frozen embryo or embryos being killed in a car or plane crash. What is to become of the embryo(s), and to whom do they belong? Under the current legislation the embryo(s) would in the UK presumably be destroyed or 'allowed to die' at the end of the agreed storage time. It would be *possible* to put forward a moral argument that we are wasting a precious human resource to no social benefit in these instances. It would be possible to thaw these embryos and use them for research which would help other infertile couples, or of course for embryo donation, which is an option on the consent form.

The Status of the Human Embryo

Our attitude to all this and to the use of human embryos for research ultimately depends on what we consider to be the status of the human embryo. At the stage freezing takes place the embryo has gone through at maximum only a few cell divisions following fusion of sperm and egg at fertilization and consists of between 1 and 8 or rarely 16 cells.

In biological terms life is a continuum being passed on from generation to generation via the sperm and oocytes, so the question is not when *life* begins, but at what point the *full human moral status* of a unique individual comes into being. The pro-life lobby on the one hand consider that the introduction of the paternal genetic material at fertilization initiates a human being with full human rights. The submission by the Nationwide Festival of Light, now incorporated into the Christian Action, Research and Education (CARE) Trust, to the Warnock Committee in 1983 states:

> the only humane ethical position (perhaps due to our present ignorance) is to treat the human embryo at any and every stage of its development with the respect due to human life at all stages.[7]

However, I do not believe we are *that* ignorant. Ignorant, yes, as to what we are really looking for as we search during development for the beginnings of something uniquely and individually the essence of a human person, but *well aware* that development is a progressive journey. The embryo differentiates into a foetus in which all the adult organs have been initiated, and these progressively grow and mature eventually into their neo-natal, infant and finally adult forms. *Biology* seems to support the gradual development with time of an organism with more and more morphological and chemical humanity. This leads to a gradualist rather than absolutist view of the emergence of the fully human individual starting from the point of entry of the paternal genetic material into the egg. Such a gradualist view has been followed from historic times, but modern biological knowledge supports the idea and gives it a new dimension (Dunstan 1974).

But does this gradualist approach mean that there are no abrupt thresholds where something of fundamental significance changes?

The potential for the parthenogenetic development of an egg has been used as evidence that no significance should be given to the point of fertilization in mammalian development. Such an argument emphasizes the continuity of life (Harris 1991). This does not in any way detract from the uniqueness of a particular combination of male and female genes, a unique new combination which produces the blueprint for a unique new individual human. Certainly in some organisms the activation of the egg can give rise to viable offspring without the addition of male chromosomes. However, there is no known case of virgin birth in mammals, although a variable degree of parthenogenetic embryonic development is possible (Kaufman 1983). Indeed, increasing evidence suggests that the passage of the gametes through the pathway of development to sperm in the testis and to the oocyte in the ovary gives rise to fundamental differences in the constitution of the genes of sperm and egg, which will effect their ability to function in the embryo and adult. This phenomenon is known as 'imprinting' (Surani and Norris 1984; McGrath and Solter 1984; Reik 1989; Monk and Surani 1990; Barlow *et al.* 1991): the chromosomes somehow 'know' whether they are derived from the male or female and react accordingly. Thus sperm plus egg cannot be considered as equivalent to a doubling of the genetic material of the egg, not only because of the introduction of new and different genetic material which occurs at fertilization but also because of chromosomal imprinting. I conclude that fertilization must be considered a threshold at which a significant event brings the ingredients together for, and initiates the pathway to, a unique human individual. I do not conclude that human personhood emerges at this point.

So is there a *definable point* at which a human foetus can actually be said to gain the status of a full human individual?

Knowledge of human development forces us to conclude that the point of full human individuality must be well before the point when it can survive outside the uterus: the foetus exhibits many well-defined human characteristics and responses in the second half of gestation, even though it cannot live outside its mother's body. Furthermore, technological advances have meant that increasingly premature babies can be kept alive *ex utero*.

This definable point has been suggested as that at which the embryo begins to be able to sense its environment via its nervous system. Consideration of full humanity has frequently arisen in debating the question of whether it is ethical to carry out research on the human embryo. In this context the point at which pain can first be experienced has been viewed as significant. Some, such as

Donald Mackay, former Director of the Research Department of Communication and Neuroscience at Keele University and an evangelical Christian, have suggested that the ability to participate in conscious neural activity or to undertake uniquely human forms of interaction in 'self-conscious' or self-aware relationship is the point at which a unique human soul arises (Mackay 1987). The Warnock Report came down in favour of a cut-off for research at fourteen days, when the primitive streak arises and the anterior–posterior embryonic axis is established. This conclusion was reached following advice from a number of eminent developmental biologists. At this time we can identify that we have a single developing individual rather than twins, which can arise by the formation of two embryonic axes at this stage. This recommendation from the Warnock Committee has been incorporated into the legislation of the 1990 Embryology Act, precluding research beyond fourteen days *in vitro*.[8]

I have already mentioned the gradualist view, and this seems a good one in the light of our knowledge of developmental biology. From this perspective we would accord the embryo progressively more rights and greater respect as it develops and comes nearer to what we recognize to be a fully functional and independent human being. Thus there are a number of criteria which must be applied in our assessment of humanity, and these must *all* be taken into consideration. In practical terms this does, however, make it difficult to define at which stages particular approaches or treatments are permissible (biopsy of cells, research procedures, etc). At some point most people would consider that we have to acknowledge, to quote G. R. Dunstan (1974; Watson and Tuddenham 1990), 'a threshold at which experiment must cease, a step which must not be crossed, for beyond it lies the life of a man' (or woman, I must add!). It seems a reasonable *interim* judgment at present to take the formation of a single individual embryonic axis leading to the initiation of the nervous system at primitive-streak formation to be such a point. Future knowledge may force us to modify this idea, and it would be irresponsible to sit back complacently, thinking that the ethical arguments are concluded with the passage of the 1990 Act.

As a Christian I conclude that the way God knew each one of us before birth *in utero*[9] may not be as He does now. We can only surmise to what degree we approached the uniquely individual 'you' or 'me' at each morphologically definable embryonic stage.

Furthermore, we cannot expect precise guidance about embryonic stages in the context of the lack of knowledge of reproductive biology of the biblical writers. God is much greater and more far-seeing than to be limited by the unpredictable combining of gametes in the IVF clinic, or in the human body for that matter. It is difficult to believe, as Paul Watson and Ted Tuddenham (1990) point out, that all the millions of embryos lost during the first month of development following natural conception are created as fully fledged human beings with full human rights only to have their lives terminated almost immediately. Many of these embryos are probably unviable because of chromosomal abnormalities, but to use this as a 'get-out clause' ignores embryos which are apparently normal and may be expelled because of maternal failure. IVF leads to the production of a few embryos, the early development of some of these and the implantation of still fewer. A small percentage of these end up as new-born babies. We are limiting God dramatically to suggest that He is not able to discern the outcome of all the fertilized eggs from the beginning and to 'know' which will grow to fully developed adult human beings and which will not.

This pre-knowledge of the potential to be reached by each embryo is only available to *us* in retrospect unless we premeditate the disposal of a 'spare' embryo. Are we then playing God when we deliberately terminate the life of such an embryo? We can use our considerable God-given intelligence and ability to make choices for good or for evil in all areas of life. Perhaps here our choice must be to take the pathway which leads to the least harm and greatest benefit to the most 'fully' human.

6 The possibility of manipulation of the embryo

Here I consider our response to techniques for the assessment of the embryo's genetic composition with possible diagnosis of genetic aberrations and consequently the potential for alteration of the genetic material by addition or deletion of DNA (gene therapy). The question of what kind of research is permissible on the human embryo, if any, must also be addressed. Our response to these issues depends on our answers to a number of questions. The question of the moral status of the human embryo is the first one to be asked. Practically, we must decide this for the embryo during the period when culture *in vitro* is currently possible and up to some later stage to which it will in the future be possible to keep the embryo developing normally outside the mother's body. I have considered this issue already, but we have to be aware that what we do to the embryo during its first cell divisions affects it later in development up to the fully fledged adult human being, should it survive that long. That influence may, of course, be as stark and fundamental as promoting its life or death.

Embryonic screening for genetic disease occurs routinely in the post-implantation phase of development (McLaren 1987). Current techniques in molecular biology allow the amplification of DNA so that it is possible to detect specific stretches of DNA in the genes of a single cell if complementary DNA probes for these stretches are available. IVF combined with the ability to remove individual cells from the early embryo without detriment to its future development allows pre-implantation diagnosis of genetic defects using the biopsied cell. We therefore have to decide if it is morally acceptable to sample embryonic cells for the purpose of diagnosing genetic diseases and then to dispose of embryos carrying defective genes.

In 1990 a group from the Hammersmith Hospital reported the first pregnancies from this technique (Handyside *et al.* 1990). They biopsied the embryos of five couples at risk from recessive X-chromosome-linked diseases where only male children carrying the defective gene would be affected. This is because in the male there is only a single X-chromosome, and if this carries the defective gene it is not balanced by a second normal gene, as would be present on the second X-chromosome of the female. This normal gene would 'neutralize' the effect of the defective gene in heterozygous female children. Thus expression of the defective gene will be apparent only in the male children.

Although no DNA probes were available for the genes of concern to the Hammersmith group, it is possible to use a probe for the male Y-chromosome and thus to detect male embryos and avoid replacing them in the uterus. In two cases where cell biopsy and DNA analysis was carried out, for couples at risk from adrenoleukodystrophy and X-linked mental retardation, female embryos were replaced after biopsy and pregnancies established. This technique avoids the alternative of screening at 8–10 weeks by chorion villus biopsy (sampling the placental tissue) with associate risk of inducing abortion, or at 16–20 weeks of gestation by amniocentesis (sampling of the amniotic fluid of the embryo). Following these tests, if the foetus is found to be carrying the defect, the parents have to decide whether to abort a foetus at a relatively late stage or continue with the pregnancy of a child with such a genetic disease. To be able to reduce the number of abortions of this type (whether or not one considers that it is morally acceptable to carry out such abortions in any case) must be seen as a considerable benefit.

These new techniques have recently been applied to the human oocyte, and it has been shown to be possible to detect the β-haemoglobin gene in single oocytes, and the 'polar body', the vestigial DNA from the cell division which gave rise to the oocyte and which is still associated with it after ovulation (Monk and Holding 1990). By sampling this cast-off DNA it would be possible to predict, in certain cases, if the oocyte itself carried the deleterious mutation where the mother carries one defective copy of the gene. This has been attempted for the mutant β-haemoglobin gene which gives rise to sickle-cell anaemia in homozygous individuals (those carrying the defect on both chromosomes of a pair). No pregnancies have so far been reported from fertilized oocytes after removing the polar bodies for analysis, but this is no doubt only a matter of time.

The possibility of manipulating an embryo with the intention of replacing it in the uterus carries the risk of its inadvertent destruction. In other instances if embryos are part of a research study there will be the *intention* of the ultimate demise of the embryo rather than its replacement.[10] This forces us to address the question: is it morally acceptable to sacrifice individual embryos for the good of future embryos, independent human beings, or society in general?

First, I do not believe that we can consider the human embryo at whatever stage in the same light as that of a mouse or even a monkey, as should be apparent from the arguments I have already outlined. Second, wherever possible, experiments to gain knowledge of use to medical science must be carried out in appropriate animal systems (mammalian cell lines or embryos of other mammals) before research on human embryos is even considered. I am aware that this is a speciesist argument and a view diametrically opposed to that of the animal rights movement. However, we know that there are differences between even the best animal models and the human in many cases. The way that human and mouse females regulate X-chromosome gene expression depending on the number of X-chromosomes in the cells (so-called dosage compensation) is one instance (Ashworth *et al.* 1991); the nutritional requirements[11] and the changing pattern of metabolism of the developing embryo are others (Hardy *et al.* 1989; Leese *et al.* 1991). This means that to find out the way that the human embryo and foetus really develops and functions and to devise medical techniques to counteract malfunction we may ultimately need to carry out limited experimental work in the human.

I should point out here that the Human Fertilization and Embryology Act does allow for research on human embryos, but only for clearly medical benefit. It permits, under licence, research which will lead to: advances in treatment of infertility; increased knowledge of congenital disease; increased knowledge of causes of miscarriage; better methods of contraception and research into better methods of diagnosing genetic abnormalities. The Act carefully separates licensing of research projects relating to human embryos from any procedures leading to the replacement of the embryo in order to establish a pregnancy. Thus any embryo which has been subjected to a research procedure cannot be replaced in the uterus.

Again our attitude to this will depend on our consideration of the status of the embryo at this time. Even if it is not a fully developed human being, it must still be treated with respect, since it is gaining further human attributes with each day of development and we must assume it will become a human individual unless we find otherwise. I would contend on the gradualist argument that our respect and caution must increase with advancing development up to the primitive-streak stage as the current absolute cut-off (although we may need to consider an earlier or later threshold in the light of our future knowledge). This means that research which may be permissible for the earliest cell-division stages might not be acceptable after, say, the stage at which the embryo would normally be implanted and with the differentiation of embryonic tissues.

To give an example, at early pre-implantation stages some *might* consider it acceptable to introduce radioactive precursors for various cell products, proteins, lipids, etc. into the embryo and carry out a biochemical analysis which would inevitably involve destroying the embryo. This information would benefit future couples and embryos by helping us to understand the normal and abnormal synthesis of these components with the aim of ameliorating problems in their production which might account for lack of development of other embryos. At later stages, approaching the primitive-streak stage, it might be acceptable only to collect and analyse secreted proteins or metabolites and to look at the removal of precursors from the medium by a non-invasive process.

CHANGING THE GENETIC MATERIAL

Theoretically the next step beyond diagnosing genetic defects must be to attempt to mend genetic damage by 'gene therapy'. At present this appears to be a grey area under the 1990 Act (Schedule 3) and is not permitted for embryos which are to be replaced for establishing pregnancy.

So is it morally acceptable to change the genetic material therapeutically, for instance by the replacement or repair of a defective gene? Should we consider gene-correction medicine any differently from therapy which corrects the defects in the protein-products of those genes? This is a very difficult question, and it is avoiding the issue to consider only what is possible now. Much of the outcry in the early and mid-1980s prompted by the IVF question stemmed from the fact that the technology had progressed much in advance of society's ability to deal with the social and ethical issues which arose from it. Perhaps we can at least learn from this and not be caught out in the same way with the next technological advances. We therefore need to increase public awareness of what may realistically become *possible*, while counteracting the views of both the alarmists and those who seem to suggest that the new techniques will lead to abolition of all genetically carried diseases.

Having said that we must not avoid the question, I must point out that at present we still cannot directly and accurately target a specific gene to a particular cell or organ in the body, although such methods are being developed (Riordan and Martin 1991). It has not yet been possible, for instance, to incorporate a 'homing' signal which would bind to a specific cell surface component, a non-degradable carrier (to allow its transport through the blood to the cells in question) and a means of enhancing up-take by the cells and integration into the correct site in a particular chromosome.

Many articles have been written on the advances and limitations of genetic engineering, and a considerable number of these have concerned themselves with the risks and the ethical issues surrounding this technology *per se* (see Fincham and Ravetz 1991; Anderson 1985), including those in this volume. One important requirement for the successful introduction of a gene into a cell is that for normal gene function the new gene needs to be in the correct position in relation to other genes and stretches of DNA which will regulate it. It has recently been shown possible to target genes into specific sites in mammalian cell lines in culture using a technique called homologous recombination (Capecchi 1989). Retrovirus has also been seen as an efficient method of introducing genes into cells, but there are problems of safety because these viruses may exchange genetic sequences with other retroviruses and rearrange their own structure, i.e. they are unstable and have the possibility of promoting malignancy. However, new safer retroviruses are being constructed (Fincham and Ravetz 1991; Anderson 1985). Experiments on somatic cell gene therapy (applied to tissue cells rather than gametes or the early embryo) in animals have been carried out most extensively using bone-marrow cells. These can conveniently be removed from the body where they can be exposed to vectors carrying the gene to be inserted, and then the cells can be replaced. In this way it has been shown that a human lymphoblast cell line homozygous for Lesch-Nyhan syndrome (in which the gene coding for the enzyme hypoxanthineguanine phosphoribosyl transferase (HPRT) is defective) could be restored to normal function by the introduction of a normal HPRT gene carried by a retrovirus (Willis *et al.* 1984). The introduction of genes into the mouse embryo by retrovirus has been tested, but in one study only 7 per cent of embryos infected both survived to birth and carried the gene in the chromosomes. The single animal analysed further had copies of the gene in a number of positions in the genetic material (Huszar *et al.* 1985). All the same, such human somatic cell therapy has now been initiated in the USA.

In the mouse, embryonic stem cell lines which maintain many features of embryonic cells can be grown in culture. The cells can be injected into the embryo just before the stage of implantation and will contribute to many of the tissues of the body, including the gametes (Robertson and Bradley 1986). Homologous recombination has been used to target genes into specific regions of the chromosomes of these cells. Most research workers have taken the approach of trying to knock out a normal gene by the introduction of a defective gene so that the influence on the cell of lack of expression of that gene can be analysed. This has been carried out quite successfully (Capecchi 1989; McMahon and Bradley 1990; Mansour *et al.* 1988). However, it has also been possible to do the reverse and use a normally functioning gene to rescue an embryonic stem cell with a defective gene (Doetschman *et al.* 1987; Thompson *et al.* 1989). This has been suggested as one route to gene therapy, since it would be theoretically possible to carry out a similar approach with human stem cell lines. Normal genes could be introduced into an embryo carrying a genetic defect via injection of human stem cells just before implantation. A futuristic picture might be the growth of a stem cell line from one embryo produced by a couple carrying the disease and the freezing of a second embryo. The normal gene could be introduced into the cell line by the most efficient method available for homologous recombination. Only a few cells will incorporate and integrate the normal gene correctly (0.5 per cent would be high by current methods (Capecchi 1989)). Those which did could be selected for and cloned. The integration could be carefully checked and cells carrying the gene correctly could be introduced into the thawed second embryo. Their incorporation into tissues could rescue the second embryo from the defect.

Is this type of approach really feasible in the near future, and if so, is it acceptable? There are major problems here. The first is the success rate, which I have already indicated is low for homologous recombination and is compounded by the fact that only a small proportion of embryos into which the cells are introduced both survive to birth and contain large numbers of corrected cells with a wide distribution. Since not all the embryonic cells are from the embryonic stem line grown in culture, the embryo will be a mixture of cells which have the genetic abnormality rectified and those which do not. In experimental animals this does not seem to be a problem in terms of repairing the functional defect as long as sufficient corrected cells occupy a particular organ or tissue. But if only a small proportion of the cells of an organ have the normal gene, then there may be insufficient cells to provide the correct function. Furthermore it is normally found that although cells in which homologous recombination occurs carry the gene at the targeted site, other cells have the introduced DNA inserted elsewhere. Therefore it would always be necessary to select for those cells in which the DNA was correctly integrated. We cannot predict the effect of the integration of the introduced gene into other unselected regions of the embryonic genome. In addition, is it ethically acceptable to modify the cells of one embryo and use its cells to correct a defect in a second? Will the status of such genetically modified individuals be any different from that of unmodified individuals? Will the individuals see themselves differently (better or worse) (Chadwick 1985)? Of course such therapy could be kept secret, but this in itself might lead to problems. The unacceptable risk of failure with the current state of the technology would preclude its use at present whatever the answer to this last question, but it has to be faced now because the efficiency of carrying out this kind of operation is improving rapidly.

The fertilized single egg cell (the zygote) or mature oocyte prior to IVF would themselves be ideal candidate cells from the technical point of view for gene therapy if such therapy were considered acceptable. The difficulties inherent in selective introduction of genes into only particular cells of the body would be avoided. The gene would be present in the germ line, being passed on via the gametes to the next generation. This would be technically efficient in that the problem defect would not have to be corrected again in the next generation. However, this very fact gives rise to considerable unease (Watson and Tuddenham 1990; Anderson 1985). There would be no unpleasant invasive procedures beyond those occurring normally in IVF, and it is possible to inject genes directly into the egg nucleus, thus avoiding some of the problems of getting the DNA to the site of integration (Allen *et al.* 1987). The injection process is difficult, but the success rate can be of the same order as the percentage of offspring from *frozen* IVF embryos (2 per cent (Allen *et al.* 1987; Brinster *et al.* 1983)). Indeed nuclear injection into other cells types seems to produce a better efficiency of integration than other methods of introducing genes (Capecchi 1989).

So the technology is available, and the question is when or whether we should attempt such genetic correction as part of human medicine. *If* it were desirable to correct genetic defects by this method, many further years of research on animal cells and zygotes including those of primates would still have to be undertaken before one could consider application of new forms of this technique to the human embryo. Whether one would want to invest so much in irreversibly modifying the structure of the genome, which must inevitably be a risky procedure, is uncertain. Of course the technique does not actually lead to the introduction of an unmodifiable genetic change, but attempts at remedial gene therapy to correct a mistake could create as many problems as the first attempt at therapy. In any case, one would envisage that this technique would only be feasible, in the near future, for the cure of diseased states caused by single gene anomalies (Lesch-Nyhan Syndrome, phenylketonuria, Tay Sachs disease, β-thalassemia, sickle-cell anaemia). This will preclude its use for many diseases where more than one gene is involved, along with environmental influences (such as cancer).

Certainly there are few who see deliberately introducing new genes which are not normally present in the human, or not in the form or numbers proposed for introduction, as ethically acceptable (Anderson 1985). This seems a recipe for disaster biologically, because the genes of any organism are interacting in fine balance with one another and the introduction of further copies, or new genes, would be most likely to upset the delicate equilibrium of gene control and expression rather than enhance the functioning of the organism in any way. New mutations would arise in the original and 'new' genes, and one would never be able to 'clean up' the human genome, even if that were desirable. *Diagnosis* of gene defects followed by embryo selection, which will be feasible for many single gene defects as the appropriate probes become available, is a real possibility in the near future. The technology will be simpler and conducive to clinical use, and it is more acceptable to those who view the life of the human embryo as something unique (and linked to a specific gene complement), even at a stage when the embryo has not reached the full moral status of the adult.

Finally I would like to emphasize that although it has been stated that a cautious approach can be taken as an excuse for doing nothing, we are nevertheless talking about human life. While every effort should be made to promote the reproductive health of each individual as in any other area of medicine, to apply the relatively unknown and little-tested molecular biological techniques for altering DNA to human embryos with the aim of creating a new human being would be irresponsible at the present time. We would have to weigh up the reliability and risks of any technological approach and consider if there was a real benefit to existing people or currently developing human individuals or, if it was deemed of value, to those to come. If the benefit was seen to outweigh the disadvantages and the changes were pronounced morally acceptable to society and the individual(s) immediately concerned (parents), this would warrant its use.

For a procedure to be morally acceptable it must have two components: that which 'society' finds acceptable (and it will produce legislation to this effect), and that which individuals find acceptable. As long as the individual is within the law defining society's ethical standards, he or she can take his or her own moral stance and chose to participate in, or abstain from, medical treatments such as those outlined above. The way ethical standards are reached is complex, but it is at our peril if we neglect to consider, in total, the weighed 'just desires' and 'right feelings' of human beings, the religious arguments and the biological, medical and technical 'facts'.

At the very least we will need exhaustive research over many years using animal models before we can reassess the situation with respect to genetic modification of the human embryo. This also seems to be the view of those establishing the 1990 Act. Acts are not immutably written in stone, and after the proper development of the appropriate techniques with exhaustive safety testing in other biological systems, gene therapy along the lines outlined here might be appropriate in certain cases. That is, if it has not been superseded by another, as yet unthought of, approach!

NOTES

I am most grateful to Dr Maureen Wood for advice and constructive criticism, particularly in the section on frozen embryos, and to Mr Paul Harper and Dr Paul Sharpe for their useful comments and constructive criticism.

1 *Report of the Committee of Inquiry into Human Fertilization and Embryology* (Chairman Dame M. Warnock, DBE) (London: HMSO, 1984); henceforth Warnock Report.
2 The Human Embryology and Fertilization Act (London: HMSO, 1990).
3 Dr Pauline Webb, BBC Radio 4 *Thought for the Day*, 13 March 1991.
4 Genesis 1:28, 2:24.
5 Church of England Board of Social Responsibility 1985; Warnock Report, pp. 42–7, 87.
6 The Human Embryology and Fertilization Act, op. cit.
7 *Human Fertilization and Embryology*. Submission by the Nationwide Festival of Light, CARE Report (London, 1983), p. 12.
8 The Human Embryology and Fertilization Act, op. cit.
9 For example Psalm 139, Isaiah 49:1.
10 I have not considered here a third scenario in which an embryo is used for research purposes and then replaced. This is not currently permissible in the UK, as already outlined, but most of the points considered still apply. There are additional problems which will depend on the type of research carried out. If the medium in which the embryo was being cultured had simply been sampled by a non-invasive technique, it might be considered more morally objectionable not to replace the embryo. All research proposals really need consideration on their own merits as to whether an embryo subjected to a particular treatment should be replaced. However, this type of more flexible regulation could be difficult to specify unambiguously and might lead to disputes in interpretation of the law. It would also be difficult to police, and such flexibility was not incorporated into the 1990 Act.

11 For instance the serum requirements of mouse and human embryos *in vitro* are different (see also Edwards 1982).

REFERENCES

Ahuja, K. (1990) 'Embryo freezing – hope for future life', *National Association for the Childless Magazine* 20.

Allen, N. D., Barton, S. C., Surani, M. A. and Reik, W. (1987) 'Production of transgenic mice', in *Mammalian Development: A Practical Approach*, ed. M. Monk, Oxford: IRL Press, pp. 217–33.

Anderson, W. F. (1985) 'Human gene therapy: scientific and ethical considerations', *J. Medicine and Philosophy* 10: 275–91.

Ashworth, A., Rastan, S., Lovell-Badge, R. and Kay, G. (1991) 'X-chromosome inactivation may explain the differentiation in viability of XO humans and mice', *Nature* 351: 406–8.

Barlow, D. P., Stoger, R., Herrman, B. G., Saito, K. and Schweifer, N. (1991) 'The mouse insulin-like growth factor type-2 receptor is imprinted and closely linked to the Tme locus', *Nature* 349: 84–7.

Brinster, R. L., Ritchie, R. E., Hammer, R. L., O'Brien, R. L., Arp, B. and Storb, U. (1983) 'Expression of microinjected immunoglobulin gene in the spleen of transgenic mice', *Nature* 306: 332–6.

Capecchi, M. R. (1989) 'Altering the genome by homologous recombination', *Science* 244: 1288–92.

Chadwick, R. F. (1985) 'The perfect baby: introduction', in *Ethics, Reproduction and Genetic Control*, ed. R. F. Chadwick, London and New York: Routledge, pp. 93–135.

Chen, C. (1986) 'Pregnancy after human oocyte cryopreservation', *Lancet*, 19 April: 884–6.

Church of England Board for Social Responsibility (1985) *Personal Origins: A Report of the Working Party on Human Fertilization and Embryology of the Board for Social Responsibility*, London: CIO Publishing.

Cohen-Overbeek, T. E., Hop, W. C. J., Den Ouden, M., Pijpers, L., Jahoda, M. G. J. and Wladimiroff, J. W. (1990) 'Spontaneous abortion rate and advanced maternal age: consequences for prenatal diagnosis', *Lancet* 336: 27–9.

Doetschman, T., Gregg, R. G., Maeda, N., Hooper, M. L., Melton, D. W., Thompson, S. and Smithies, O. (1987) 'Targeted correction of a mutant HPRT gene in mouse embryonic stem cells', *Nature* 330: 576–8.

Dunstan, G. R. (1974) *The Artifice of Ethics*, London: SCM Press Ltd, pp. 66–74.

Edwards, R. G. (1980) *Conception in the Human Female*, London and New York: Academic Press, pp. 980–3.

— (1982a) 'The case for studying human embryos and their constituent tissues *in vitro*', in *Human Conception in Vitro*, ed. R. G. Edwards and J. M. Purdy, New York and London: Academic Press, pp. 371–88.

— (1982b) 'Discussion of the ethics of fertilization *in vitro*', in *Human Conception in Vitro*, ed. R. G. Edwards and J. M. Purdy, New York and London: Academic Press, pp. 359–70.

Fincham, J. R. S. and Ravetz, J. R. (1991) *Genetically Engineered Organisms: Benefits and Risks*, Milton Keynes: Open University Press, pp. 121–42.

Handyside, A. H., Kontogianni, E. H., Hardy, K. and Winston, R. M. L. (1990) 'Pregnancies from biopsied human pre-implantation embryos sexed by Y-specific DNA amplification', *Nature* 344: 768–70.

Hardy, K., Hooper, M. A. K., Handyside, A. H., Rutherford, A. J., Winston, R. M. L. and Leese, H. J. (1989) 'Non-invasive measurement of glucose and pyruvate uptake by individual human oocytes and pre-implantation embryos', *Human Reprod.* 4: 188–91.

Harris, J. (1992) *Wonderwoman and Superman: The Ethics of Human Biotechnology*, Oxford: Oxford University Press, p. 32.

Huszar, D., Balling, R., Kothary, R., Magli, C. M., Huzumi, N., Rossant, J. and Bernstein, A. (1985) 'Insertion of a bacterial gene into the mouse germ line using an infectious retrovirus vector', *Proc. Natl. Acad. Sci. USA*, 82: 8587–91.

Johnson, M. H. and Pickering, S. J. (1987) 'The effect of dimethyl sulphoxide on the microtubular system of the mouse oocyte', *Development* 100: 313–24.

Kaufman, M. (1983) *Early Embryonic Development: Parthenogenetic Studies*, Cambridge: Cambridge University Press, pp. 111–19, 145–53.

Leese, H. J. Humpherson, P. G., Hardy, K., Hooper, M. A. K., Winston, R. M. L. and Handyside, A. H. (1991) 'Profiles of hypoxanthine-guanine phosphoribosyl transferase and adenine phosphoribosyl transferase activities measured in single pre-implanatation human embryos by high performance liquid chromatography', *J. Reprod. Fertil.* 91: 197–202.

McGrath, J. and Solter, D. (1984) 'Completion of mouse embryogenesis requires both the maternal and paternal genomes', *Cell* 37: 179–83.

MacKay, D. M. (1987) *Human Science and Human Dignity*, London: Hodder & Stoughton, p. 100.

McLaren, A. (1987) 'Can we diagnose genetic disease in pre-embryos?', *New Scientist*, 10 December: 42–7.

McMahon, A. P. and Bradley, A. (1990) 'The Wnt-1 (int-1) proto-oncogene is required for development of a large region of the mouse brain', *Cell* 62: 1073–85.

Magistrini, M. and Szollosi, D. (1980) 'Effect of cold and of isoproprylN-phenylcarbamate on the second meiotic spindle of mouse oocytes', *Eur. J. Cell Biol.* 22: 699–707.

Mansour, S. L., Thomas, K. R. and Capecchi, M. R. (1988) 'Disruption of the proto-oncogene int-2 in mouse embryo-derived stem cells: a general strategy for targeting mutations to non-selectable genes', *Nature* 336: 348–52.

Monk, M. and Holding, C. (1990) 'Amplification of a B-haemoglobin sequence in individual human oocytes and polar bodies', *Lancet* 335: 985–8.

Monk, M. and Surani, A. (1990) *Genomic Imprinting. Development* Suppl. 1990, Cambridge: Company of Biologists, p. 153.

Pickering, S. J. and Johnson, M. H. (1987) 'The influence of cooling on the organization of the meiotic spindle of the mouse oocyte', *Human*

Reprod. 2: 207–16.

Reik, W. (1989) 'Genomic imprinting and genetic disorders in man', *Trends in Genetics* 5: 331–7.

Riordan, M. L. and Martin, J. C. (1991) 'Oligonudeotide-based therapeutics', *Nature* 350: 442–3.

Robertson, E. J. and Bradley, A. (1986) 'Production of permanent cell lines from early embryos and their use in studying development problems', in *Experimental Approaches to Mammalian Embryonic Development*, ed. J. Rossant and R. A. Pedersen, Cambridge: Cambridge University Press, pp. 475–508.

Smeirteghem, A. C. Van and Abbeel, Van den E. (1990) 'World results of human embryo cryopreservation', in *Assisted Procreation*, New York: Plenum Publ. Corp., pp. 1–11.

Surani, M. A. H., Barton, S. C. and Norris, M. L. (1984) 'Development of reconstituted mouse eggs suggests imprinting of the genome during gametogenesis', *Nature* 308: 548–50.

Thompson, S., Clarke, A. R., Pow, A. M., Hooper, M. L. and Melton, D. W. (1989) 'Germ line transmission and expression of a corrected HPRT gene produced by gene targeting in embryonic stem cells', *Cell* 56: 313–21.

Uein, J. F. H. Van, Siebzehnrubl, E. R., Schuh, B., Koch, R., Trotnow, S. and Lang, N. (1987) 'Birth after cryopreservation of unfertilized oocytes', *Lancet*, 28 March: 752–3.

Watson, P. and Tuddenham, T. (1990) 'A human responsibility', *Third Way*, June 18–20.

Willis, R. C., Jolly, D. J., Miller, A. D., Plent, M. M., Esty, A. C., Anderson, P. J., Chang, H.-C., Jones, O. W., Seegmiller, J. E. and Friedmann, T. (1984) 'Partial phenotypic correction of human Lesch-Nyhan (hypoxanthine-guanine phosphoribosyl transferase-deficient) lymphoblasts with a transmissible retroviral vector', *J. Biol. Chem.* 259: 7842–9.

7

MANIPULATION OF THE GERM-LINE

Towards elimination of major infectious diseases?

Janice Wood-Harper

INTRODUCTION

The feasibility of any new medical practice can never alone dictate its ethical permissibility. It is envisaged that genetic research will enable previously inconceivable and revolutionary procedures, such as the conferring of resistance to major infectious diseases through manipulation of the germ-line, to become feasible, and early deliberation on anticipated ethical problems in determination of its moral defensibility should be paramount in advance of implementation.

The Human Genome Project is being undertaken as an international co-operative effort. The US is the major financial stakeholder, with the CIS, Japan, Britain and Italy being the other main contributors. Its ultimate goal is to generate information which will augment understanding of the functional basis of human existence and, by explaining the role of genetic factors in diseases which adversely affect the lives of millions, to provide large benefits for human health through improved methods of detection, prevention and treatment. These ends may be achieved through 'gene therapy', an application of genetic engineering whereby genes are inserted into the genome to correct deleterious effects of a genetic disorder. Such therapy is already feasible for some diseases, and with more detailed knowledge of the genome together with advanced research and technology its potential is likely to be further realized.

Of the two main types of gene therapy, 'somatic cell therapy' and 'germ-line therapy', the former is more advanced. It involves genetic modification of already differentiated body cells, thus affecting solely the individual being treated, is technically the more simple and ethically the least controversial, as it raises no new ethical problems any different from those emanating from other similar medical therapies such as organ transplantation. 'Germ-line gene therapy', however, presents a fundamentally different and potentially more dangerous option. By inserting engineered genes into the germ cells, it alters not only the treated individual's genotype but also the genetic constitution of subsequent generations.

The consequences of this difference are crucial to the debate on whether it is ethical to transgress the divide between research on germ-line gene therapy and its clinical practice. General opinion is that germ-line gene therapy should not at present be contemplated[1] on the grounds of its technical non-feasibility and ethical inadmissibility. However, it may provide the only effective means of correcting certain defective genes such as those which cause multi-organ disorders, and should not be dismissed as a form of beneficial therapy solely on the grounds of current insufficiency of knowledge and research, particularly in relation to the stability of engineered genes over several generations, the potential severity of the consequences of any errors or unanticipated, deleterious side-effects.

It is predictably only a matter of time before the feasibility of germ-line therapy is demonstrated, and discussion about the ethical implications, particularly those previously not encountered in medical practice, should be taking place now. The following discussion focuses on the, as yet hypothetical, procedure of genetically engineering resistance to major infections into the human embryo, assuming that it will become feasible.

ENGINEERING RESISTANCE TO MAJOR INFECTIONS INTO THE GERM-LINE?

This procedure, in deviating from the accepted concept of gene therapy as a means of correcting a genetic disorder, may be viewed as representing a form of 'enhancement engineering' to which even more stringent precautions are relevant before granting its ethical permissibility.

Gene therapy or enhancement?

The aim of enhancement genetic engineering is to produce a desirable trait by selectively altering the genome to augment an existing characteristic. The inclusion of engineering disease resistance within this category is dependent upon whether or not the absence of resistance to a particular disease is interpreted as constituting a 'genetic disorder' for which therapy might justifiably be indicated.

The level of individual immunity to a particular disease is dependent not only upon genetic factors but also upon their interaction with environmental influences and any prior exposure to the causative antigen. No individual exhibits immunity to every disease, and therefore, if disease susceptibility is to be considered as a 'disorder', it is a universal one and all could conceivably stand to benefit to some extent from its correction. However, in circumstances where substantial positive medical benefits could be gained, particularly in relation to serious, life-threatening diseases as opposed to more minor illnesses, the argument for engineering resistance would seem to be more defensible than, for instance, enhancement of a genetic trait such as height. It appears, then, that genetic engineering of disease resistance may fall between gene therapy and enhancement engineering, representing an example where 'the boundary between a defective state and normality may be more blurred' (Glover 1984: 31) and as such poses even more serious and disturbing ethical problems than germ-line therapy.

Before addressing the problems which may arise from the procedure, it is helpful to consider briefly how such resistance may be integrated into the genome and subsequently for which diseases it might be contemplated and possible advantages to be gained.

How disease resistance might be accomplished

To engineer resistance to infectious diseases, alterations must be made to the genetic machinery which regulates the immunological system, the aim being to strengthen the body's immune response to the antigens carried by pathenogenic agents. This would necessitate the insertion of new genes or modification of existing ones such that the 'blueprint' for production of specific antibodies became encoded within the DNA. Because of the specificity of the antigen-antibody immune response, it would only be possible to confer resistance to a limited range of diseases, and therefore decisions as to which particular diseases might be justifiably selected would need to be made.

For which major infections might it be contemplated?

Infectious diseases are the major cause world-wide of premature death, particularly within the most vulnerable social groups. Because of the diverse nature of infectious disease, criteria which may operate in the choice of those diseases for which genetic treatment may be contemplated should be evaluated, and properties attributed to a particular infection in order that it be designated a 'major' disease be defined.

Diseases vary in their severity and may be classified according to mortality and morbidity considerations. Thus decisions may be based on the incidence of a disease within a population and the probability of its causing premature death. Alternatively, account may be taken of the longevity of a disease and the degree of physical or mental suffering to be expected. Whereas it would be relatively uncomplicated to distinguish between those diseases which inevitably cause acute pain or disablement to all affected patients and more minor diseases which produce mere cultural discomfort, difficulties may be expected in assigning those diseases which produce symptoms of intermediate severity to the category for treatment, and some sort of line would need to be drawn.

Also relevant would be a consideration of existing preventative measures or alternative therapies for any particular disease. This is especially relevant to diseases for which there are no vaccines currently available or any effective treatment. Viral infections, for instance, present problems for vaccine development and are unresponsive to antibiotics. Human immunodeficiency virus (HIV) infection, which is becoming increasingly widespread globally, may develop into AIDS, for which, at present, only palliative measures exist. It might therefore justifiably be chosen for engineering resistance.

Measures other than of a strictly medical nature exist for preventing the spread of infection. These include control of pathogen transmission by education and improving standards of sanitation and hygiene. Theoretically these could be best utilized in developing countries, where major infectious diseases like malaria, schistosomiasis, cholera and typhoid are endemic, but in practice complex difficulties arise based upon cultural and political differences and scarcity of resources. Similar problems could also be anticipated in the clinical application of genetic engineering techniques in such countries, an aspect which will be later discussed.

Concurrent with decisions as to which particular immunodeficiencies might be considered for correction is the requirement to evaluate possible advantages to be gained by implementation of the new procedure over any existing forms of therapy.

Possible benefits to be gained

Benefits envisaged as inherent in genetically engineering disease resistance are twofold: the alleviation of human suffering and death with the ultimate elimination of major diseases and overall economic advantages.

Elimination of serious infectious diseases

The technique might constitute, in reductionist terms, a one-time treatment, its purpose being to promote the inheritability of disease resistance and, if implemented on a mass scale, the total eradication of selected major diseases. It becomes necessary, however, to examine the validity of such reasons put forward in justification of manipulation of the germ-line.

Although it could be upheld as producing a moral good in preventing disease, it is possible that a similar end may be achieved by means which carry lesser risks of harm. In practice, somatic cell gene therapy on antibody-producing tissue may be as effective and safer. Once it becomes possible to detect an immunological defect in an individual, particularly if widespread genetic screening becomes available, then resistance in subsequent generations might be accomplished by use of pre-natal diagnosis so that defective children might be aborted. This would, however, inevitably result in an increase in terminations and the dilemma of whether disease susceptibility could provide ethically justifiable grounds for abortion.

The efficacy of such proposed genetic alteration of the germ-line should also be considered. Unlike gene therapy for a disease such as cystic fibrosis for which permanent elimination could be possible, it is doubtful whether the same beneficial outcome could be achieved in relation to disease resistance. Over time new strains of pathogen with different antigens, to which the body is no longer immune, often arise through mutation and contribute to resurgence of the disease, as has occurred in gonorrhoea. Similarly difficulties arise in effecting a cure for or producing a totally protective vaccine against the various strains of the influenza virus. Malaria, another disease for which genetically engineered resistance might be contemplated, is caused by a complexity of antigen function in the body, so that development of an effective vaccine has been elusory. These examples, which illustrate some intricacies and current inadequacies in the understanding of the immunological system, serve to question whether the anticipated benefits of the procedure could ever be manifested.

Economic advantages

Genetic engineering of disease resistance through the germ-line might conceivably result in various economic benefits. The one-time cost of the procedure would need to be weighed against the cost of repeated hospitalization or mass vaccination programmes in successive generations. Despite probable high initial investment in technology and training, it might prove to be a more cost-effective use of scarce medical resources in the longer term. Financial benefits may also be accrued by employers and the government. A healthier workforce would ensure that fewer working days were lost through occupational infectious diseases, with resultant saving in sick pay and state benefits.

Assuming the demonstrability of real benefits to be gained, it then becomes imperative to speculate upon the ethical implications, particularly those previously not encountered, before any final decision is reached to initiate human experimentation.

WOULD IT BE ETHICALLY PERMISSIBLE?

Harm could result not only to the physical health of present and future generations but as a consequence of the procedure itself or its implementation being contrary to basic ethical principles. Thus the moral defensibility of the proposed treatment must be examined. Initially it is pertinent to address the problem of responsibility for policy decision-making.

Who will decide?

Complexities may be contemplated in decisions on issues such as the provision of services and selection of individuals for treatment.

> The question of who decides the application of technology in individual cases must be addressed, whether it be individual genetic counsellors, codes of practice, legally established regulatory committees or parents, and whether it is freely available to all or only to those who can pay, or only to those judged to be at 'significant risk'.
>
> (Macer 1991: 205)

In the UK any provision on the NHS would necessarily involve government resource allocation. Alternatively private clinics might well dictate numbers to be treated and accessibility to facilities. It is to be expected that several factors would contribute to the final decision to treat, but the question of who is best qualified to decide which genes should be possessed by a future generation, whether it be the government, medical profession or parents, poses inherent ethical problems (Glover *et al.* 1989: 140), including, importantly, consent to treatment.

Consent to genetic selection of the next generation

The embryo, although it can neither participate in decision-making nor consent to treatment which would permanently alter its genotype, possesses rights which should not be violated. The ethical defensibility of consent by parents on behalf of the embryo to a procedure which may have unforeseen, deleterious repercussions, not only for the child-to-be but for its descendants, deserves careful consideration.

Under normal circumstances, the unique (except for identical siblings) genotype inherited by an individual can neither be predicted nor requested. Genetic manipulation of the embryo would enable parental pre-determination of desirable traits for their children, and it is arguable whether this should be permitted in view of possible unexpected resultant harm and disrespect of the rights of subsequent generations who may wish that they had not had their genes tampered with. Conversely, if such inheritable genetic alteration increases longevity by removing the threats of a life-threatening disease, then it may be perceived as a morally responsible action.

Consent also raises the problems of communication and comprehension of complex genetic information so that the parents are adequately informed on the procedure and its risks and of possible parental self-interest. Macer (1991: 206) suggests that any benefits, rather than being viewed primarily as advantages for the next generation of genetically selected individuals, may be attributed to parents who, by having their embryos screened and altered, will avoid the expenditure of future time and costs in caring for diseased children.

Genetic screening

Future involvement of a large proportion or even the majority of the population in genetic screening is anticipated once information and technology are available. Although it is generally viewed that such screening should be voluntary rather than mandatory, this cannot necessarily be ensured as institutions will have increasing financial interests. The coercion of patients to undergo screening cannot be ruled out, and the implications for the medical profession, public policy, employers and insurance companies could be extensive. The overall costs could be prohibitive, but

> on the other hand, extensive screening programs might be justified by the knowledge that disease prevention can and will save immense amounts of money that otherwise would be spent on treatment.
>
> (Friedmann 1990: 412)

Recorded information, as a result of screening, about those individuals who have been genetically engineered for resistance to infectious diseases may create problems as to its confidentiality and availability to interested parties.

Access of information to third parties

It is probable that initially relatively few individuals would be genetically altered by the new procedure, and deliberation would be necessary on whether full information should be made available to the public via the media or anonymity be required for the protection of children not yet capable of expressing their views. With predictable expansion of genetic diagnosis and more widespread practice, confidentiality could be increasingly difficult to maintain. There may ultimately be a need for restricting development of genetic knowledge or for examining and restructuring socio-political institutions in order to avoid the possibility of such information being used 'as a tool for injustice and inequity' (Friedmann 1990: 412). Legislation could provide a means for guarding against discrimination resulting from both the misuse of genetic information and obligatory disclosure to third parties by enforcing strict regulations for storage, retrieval and use of genetic data.

> The notion of relying on the principle of 'medical confidentiality' has already proved a flimsy barrier to those who claim a right to know – be they employers, insurance companies or the police. Those beavering away at unravelling the secrets of the human genome have, by and large, yet to take on board the implications this knowledge will bring.
>
> (Joyce 1990: 55)

Employers or insurance companies are likely to have substantial interest in gaining access to information on individual genomes. Whereas employment or insurance prospects for resistant individuals may be improved, non-treated people may be subject to discrimination if genetic information is freely available. Insurers may impose higher premiums on or deny coverage to such individuals, and companies employing workers in occupations where there was an above average risk of contracting a particular infectious disease may select their employees on the basis of their genetic credentials in order to avoid occupational lawsuits (Joyce 1990: 53) or costs in improving the safety of workplaces. Policies involving routine, or even mandatory, screening and monitoring would seriously undermine the autonomy and restrict the freedom of individuals.

Cooper and Barefoot (1987: 51) oppose the concepts of mandatory genetic testing and offers of reduced premiums as incentives for inducing individuals to be tested for disease susceptibility. Also Wertz *et al.* (1990: 1212), from their study of responses of medical geneticists in eighteen countries, found that, while there was agreement that 'third parties, such as health insurance companies and employers, should not have access to personal genetic data unless consent has been obtained', difficulties were anticipated in countries where there is private health insurance and the possible need for extensive legal protection acknowledged.

As well as expectations of a gradual erosion of autonomy with respect to restrictions in employment, individual freedom of choice could be encroached upon in other spheres as the result of alteration of genotype. Individuals might eventually be allocated, by various institutions, a social role in life based on their genetic constitution, and 'personal reproductive decisions will need more serious consideration in the future' (Macer 1991: 204).

Individual liberty in reproductive decisions

In order for the advantages of genetic modification to the germ-line to be perpetuated in future generations, it will be necessary to ensure that the new beneficial genes are transmitted effectively by inflicting controls on choice of reproductive partners. Harris (1992: 15–17) proposes that a new breed of transgenic humans, possessing functions enhanced by genetic engineering, would be created and restricted to breeding exclusively with similar transgenic individuals. This may lead to an increase in consanguinity or inbreeding among treated, related individuals, initially among the 'privileged few' who would be severely restricted in their choice of partners. Inbreeding itself would be likely to effect an increase in incidence of a variety of genetic disorders and so negate any benefits previously accrued. It is also possible that such controls on breeding could be enforced in order to protect any financial investment by the government. As more types of genetic alteration became commonplace, so individual freedom in the choice of a genetically compatible partner would become even more limited and prone to complexities such that 'mating agencies' holding comprehensive details of personal genetic profiles might become necessary, raising additional problems associated with confidentiality and access to genetic data.

The following issues centre upon the problems inherent in resource allocation such that all those who stand to benefit have equal opportunity of access and in the just and equitable identification of target populations.

> If people's lives and fundamental interests are of equal value then it is unjust to treat people differently in ways which effectively accord different values to their lives or fundamental interests. . . . Where literally all cannot be benefited, equality

> requires that the method of selecting who will benefit and who will not is fair.
>
> (Harris 1992: 193)

The ethical principle of justice in relation to the allocation of genetic treatment and its consequences for those who would be treated can be considered both on an individual and on an international basis.

Allocation of resources

The technology involved in genetic engineering will be complicated and, at least initially, limited and expensive. It will necessitate the training and regulation of health-care providers knowledgeable in genetics and the use of the new equipment. Consideration must be given as to the most efficacious use of scarce resources and the prioritization of needs; whether the overall benefits to be gained in effecting long-term immunity with possible future savings in ongoing costly hospitalization of diseased patients justify the initial financial investment at the expense of other current therapies.

Alternatively susceptibility to disease might be viewed, like infertility, as not strictly an illness and therefore not to be given high priority by the NHS. If this were so, then treatment might be restricted almost entirely to the private sector, a situation which would present serious ethical problems in terms of equality of access to treatment. It may also be expected that people would not initially be clamouring after the new treatment and, by the time that benefits were acknowledged, technology might have progressed to the point where commercially produced, inexpensive self-treatment kits became available for use by couples planning to procreate. This, however, may be speculating too far into the unknown, and for the moment it remains sufficient to attempt to foresee those problems which could result from probable inequalities in access to treatment.

Justice between individuals

If just or fair means are to be employed in the allocation of resources, then all individuals who could potentially benefit and would choose to do so, or in this case whose parents decide on their behalf that it is in their interests, would have equal opportunity of access to treatment.

It is doubtful, however, whether this ideal could ever be realized in regard to genetic correction in view of the unlikelihood of any general provision of treatment. Therefore, rather than selection criteria being formulated with a view to promoting justice, unequal distribution of services and possible spread of private, for-profit clinics would give privileged access to those who could pay (Wertz *et al.* 1990: 1212). Even if insurance cover were to enable forward-planning prospective parents to amass funding for the treatment of their future progeny, access would still be restricted on the basis of financial status, which is an unjust means of apportioning resources.

The advantages to be gained in society by genetically engineered people could be immense, and resultant discriminatory practices between the treated and the vulnerable who had not received protective genes would mean that allocation of treatment based on any but ethically sound grounds of medical need, which, for example, might include those most at risk or unable to receive immunization through vaccination, rather than on personal wishes, finances or comparative social worth, would be unjust. Discrimination could, however, work in the opposite direction. French Anderson proposes that 'fortunate' individuals who received the genes might be discriminated against by those who were left out, and, in citing an Office of Technology Assessment (US Congress, 1987) which indicated that society would respond positively towards an individual who received a gene in order to correct a serious disease, comments that

> it is quite another matter to know what our society's attitude would be towards persons who used genetic engineering to try to make themselves or their children 'better' than normal.
>
> (Anderson 1989: 689)

Overall it would seem likely that access to treatment for individuals would violate the principle of justice and promote inequalities between individuals. On a larger scale, similar injustices might operate between nations, particularly between developed and undeveloped countries.

Justice between nations

There is concern as to whether those nations with the largest investment in the Genome Project necessarily will or should be those which benefit most from the new knowledge. Will a data-sharing ethic be upheld or, in view of the fact that human genetics has always been intensely competitive (Roberts 1990: 953), will information be withheld from countries which have not supported equivalent research projects? For example, tensions already exist and could be compounded surrounding American doubts about the continued stature of the US as an economic superpower and its biotechnical dominance over Japan. Galloway acknowledges the probability of an international bargaining table by suggesting that

> whoever gets the human genome data first will decide what will happen to them, and will be in an unassailable position to dictate terms over its commercial, including its medical, exploitation.
>
> (Galloway 1990: 41)

Many view the information contained in the human genome as belonging to the world's people, rather than to its nations (Watson 1990: 48); that 'society's rights should transcend proprietary rights . . . (and) the claims that we can all make upon the genome should make the shared authorship and ownership legally compelling' (Macer 1991: 211). The implication is that there should be international co-operation to promote equal access to data and opportunity to benefit, especially those who have greatest need. As Rawls writes,

> the saving of the less favored need not be done by their taking an active part in the investment process. Rather it normally consists of their approving of the economic and other arrangements necessary for the appropriate accumulation. Saving is achieved by accepting as a political judgment those policies designed to improve the standard of life of later generations of the least advantaged, thereby abstaining from the immediate gains which are available.
>
> (Rawls 1972: 292–3)

Developing countries support vast populations most in need of the new technology. They cannot contribute resources to research, nor can they mount the professional expertise necessary for implementation, yet if there is to be justice between nations, they should be entitled to benefit to the same extent as those which are more affluent. It is doubtful, even with altruistic motivation on the part of developed countries through financial support, whether the practice of genetic engineering could achieve its intended beneficial effects in substantially reducing disease. Difficulties, similar to those experienced in attempts to introduce contraceptive methods, can be expected arising from cultural differences such that 'the introduction of genetic protection will inevitably widen the gulf between the industrialized world and the non-industrialized world' (Harris 1992: 197).

Inequality may also arise between nations during the experimental phase of the new procedure. Genetic engineering of the germline, in digressing markedly from ordinary medical practice, might, in its initial clinical application, constitute research rather than innovative practice (*Report of the Committee of the Ethics of Gene Therapy* 1992: 11–12). This raises the question of whether vulnerability, particularly of those in undeveloped countries, might be exploited by the coercion of vulnerable people in early human trials.

To summarize, it would seem unavoidable that inequalities between individuals and nations would be a consequence of clinical implementation of the new technique. Although justice may be viewed as a moral ideal for which to aim, in the context of access to medical treatment it is difficult to achieve. Current medical services, particularly on a global scale, are seldom perceived as being allocated on a just basis, but yet other overriding moral considerations provide justification for their practice.

A frequently expressed fear is that, were the practice to be ethically sanctioned, temptation to push the boundary beyond the limits of moral acceptability would be inevitable.

Dangers of the slippery slope

Genetic intervention has the potential to be eugenic, that is, to 'improve' inborn traits to the ultimate advantage of the race. The categorization of genetic engineering of resistance to infectious disease as an ethically permissible technique falling within an indistinct divide between negative and positive eugenics might promote the danger of the slippery slope into enhancement engineering, which is generally viewed with such apprehension as to be beyond consideration. Anderson (1989: 689) points to the possibility of any demarcation shifting, as procedures become more commonplace, so that we might 'slide into a new age of eugenic thinking by starting with small "improvements"', and for this reason advocates a delineation based on medical and ethical criteria. Glover (1984: 30–3) similarly suggests that a distinction be made according to what is morally acceptable.

> The issues of germline therapy and enhancement engineering need to be debated widely in society, but arguments that genetic engineering might someday be misused do not justify the needless perpetuation of human suffering that would result from an unnecessary delay in the clinical application of this potentially powerful therapeutic procedure.
>
> (Anderson 1984: 408)

Dawson and Singer (1990: 485) acknowledge that with more information on the human genome, including identification of individual disease susceptibilities, there is a risk of some undesirable form of eugenics becoming a social norm. However, they also propose that not all eugenic policies are undesirable and, because the potential for eugenics already exists, their introduction would be socially or politically determined, rather than being dependent on genetic knowledge and technology.

A more direct and predictable effect of genetic engineering on the germ-line will be changes to the human gene pool, the sum total of all human genes, with resultant consequences for humanity.

Consequences of changing the human gene pool

It is proposed that intentional alterations in the fundamental nature of our existence resulting from interference with the gene pool will change our 'humanness'. A resolution (Budiansky 1983: 563) based on mainly religious signatories and released in the US, calling for an immediate ban on germ-line therapy, cited 'new advances in genetic engineering technology' that 'raise the possibility of altering the human species', and concluded that humans did not have the right to alter the human gene pool by deciding which genetic traits should be introduced and which eliminated.

The insertion of one or two genes is unlikely to produce a significant effect on the human gene pool, but if genetic alteration of the germ-line were practised on a mass scale, permanent changes could reduce the genetic variability on which evolution is dependent. It is doubtful, however, whether this alone could justify prohibition of such practices. Genetic engineering for disease resistance would constitute selection for a desirable trait and as such can arguably be no different from the indirect effects of, for example, exercising a choice of mating partner, genetic counselling or pre-natal screening, which are generally considered to be morally acceptable unless objections are mounted on the grounds of undesirable consequences such as abortion (Anderson 1989: 682–3; Glover 1984: 30–1).

A major concern is expectation of the danger of irreversible, undesired consequences resulting from ignorance or design. However, other medical therapies regarded as being ethically defensible, like radiotherapy, which can produce genetic damage, or those which successfully treat genetic diseases such as diabetes, haemophilia or immune deficiency and thus preserve deleterious genes that would otherwise be deleted, also indirectly contaminate the gene pool and increase the need for considerable future expenditure on treatment of the genetically infirm (Motulsky 1983: 136). Yoxen proposes that

> one eugenist view might be that somatic gene therapy suspends the action of natural selection against genetically inferior individuals, who pass on their genes, when otherwise they might have died, to the detriment of the human race.
>
> (Yoxen 1986: 167)

This argument against somatic gene therapy could be advanced in favour of alteration of the germ-line as a means of ensuring that only 'beneficial' genes are inherited by future generations and therefore of improving evolutionary prospects by artificial selection. Such interference with the identity of the species and the natural process of evolution may be construed as an attempt to 'play God', and its moral permissibility warrants careful consideration.

Is it permissible to 'play God'?

Many religions take the view that we are made in the image of God and therefore that the germ-line should remain sacrosanct. Zohar (1991: 275–89) proposes that an unaltered, intact genotype is essential for maintaining personal identity at the embryonic stage and that its uniqueness constitutes the essential nature of each individual. He therefore contends that genetic alterations, in removing deficiencies, may be viewed as 'not curing the existing embryo, but using it as raw material for producing another (superior) one'. Thus objections to 'playing God' centre on the concept that we should allow 'natural' mixing of genes and have no right to intervene by altering the course of nature.

This position, however, takes no account of the unique human abilities made possible by a highly evolved brain. Just as human beings are able to impose undesirable changes on the environment or make choices about their way of life or social habits, factors which frequently increase the spread of disease, so they have also acquired the means of developing the knowledge and technology necessary for 'improving' the species. 'Natural' processes have allowed disadvantageous traits to develop, and prior to advanced medical care many adversely affected individuals unsuited to the environment of the time or not amenable to existence did not survive to reproduce. It could be argued that, if such premature deaths are preventable, both in the present and in future generations, then a moral obligation might operate such that it would be unethical not to allow vulnerable people to benefit.

An argument against 'playing God' is that intervention would produce deleterious effects on the delicate balance of nature. Disease can be viewed as a form of natural population control, and engineering resistance into the germ-line may result in overpopulation, particularly in developing countries, where more would die of starvation. Although this is a valid argument, it cannot provide ethical justification for such people being denied treatment and, in practice, a higher child survival rate might effect a greater willingness to limit family size.

Finally, natural selection proceeds very slowly and, at a time when we are causing rapid changes in our environment, it is conceivable that we will need to effect similar rapid inheritable alterations to our genotypes for protection against increasingly hostile surroundings. Thus artificial acceleration of the evolutionary process and even forms of enhancement engineering which currently may seem 'frivolous' or morally indefensible, for example to increase melanin production in the skin to afford protection from harmful effects of ozone depletion, might be justified to ensure the survival of the species.

To summarize, the particularly complex and unique ethical issues involved in genetically engineering resistance to disease through the germ-line necessitate that decisions as to its moral permissibility be made after timely deliberation before its inception based on the implications not only for the embryos which are corrected and the people they will become but also for future generations. The possible benefits to be gained by individuals and humanity must be weighed against the probable sacrifice of justice and any risk of harm. If, finally, it is accepted as ethical, then stringent controls over its implementation should be formulated to minimize risks arising from either a lack of knowledge or its misuse.

CONTROLS WHICH WOULD BE DESIRABLE

Such controls could be envisaged as twofold: those which would operate in advance of initial clinical implementation of the procedure and those which would make provision for the authorization of specific applications and promotion of responsibility for regulating and monitoring their progress.

Prior to implementation

Before the introduction of the proposed programme of genetic engineering, it would be essential to ensure that certain standards were met regarding its effectiveness and safety and that its practice by skilled professionals was concurrent with public awareness and approval.

Assessment of effectiveness, safety and costs

Initially the effectiveness of engineered genes in producing inheritable disease resistance and safety of their transmission to subsequent generations would have to be demonstrated. This would necessitate prior experience of somatic cell gene therapy in practice and extensive animal trials on the effects of manipulation of the germ-line over many generations in advance of human experimentation. Decisions as to the efficacy of the procedure would be dependent upon an assessment of the inherent risk. A favourable ratio of benefit against risk, indicating that the probable benefits to be gained were expected to exceed the possible harms, would be required. Identification of acceptable levels of risk would vary. The severity of the particular disease, individual susceptibility or expectation of success may provide justification for a potentially considerable risk (Anderson and Fletcher 1980: 1294). Ultimately it would be imperative to protect against the dangers of premature clinical application before the above requirements were satisfied.

Assessment of cost-effectiveness should be based on comparison with any existing therapies for a particular disease and the provision of new technology, including that needed for genetic screening, together with the costs of maintaining records and providing a monitoring system (*Report of the Committee on the Ethics of Gene Therapy* 1992: 14), both of which would be essential in this case, where genetic modification would affect subsequent generations. Additional costs would be incurred in educating medical professionals. The widespread practice of genetic engineering would require not only a high degree of technical expertise but also counselling skills based upon a substantive knowledge of new genetic developments and awareness of all possible consequences of patient choices so that information could be communicated effectively.

Whereas some of the above considerations would apply equally to any other proposal for implementation of a new medical procedure, genetic engineering of the germ-line would also necessitate, in view of its wide social implications, both public awareness and approval.

Public awareness and approval

Public fears about the potential abuse of the new genetic technology are, at least in part, based upon the extremes of popular fiction. The need for education at all levels is recognized by Motulsky (1983: 140), who, in acknowledging agreement in most societies that completely novel medical procedures require public discussion and that scientists should be accountable to the public prior to their inception, proposes the need for improved formal and informal education in human biology and genetics and responsible reporting of new developments by the media.

Anderson (1985: 286–7), in contrasting germ-line therapy with somatic therapy and other controversial procedures, such as *in vitro* fertilization and holistic treatment of cancer, explains that, whereas prior public approval is not being specifically sought for the latter, which can proceed on the decision of the patient, the impact of germ-line therapy goes beyond the individual in affecting future generations and society as a whole. He therefore advocates the importance of assent as indicated by an informed public prior to the initiation of clinical trials. Thus the need for public understanding and approval of clear objectives and probable benefits is essential in order 'to preserve the right of society to determine how the achievements of science are used' (C. M. Clothier, foreword to *Report of the Committee on the Ethics of Gene Therapy* 1992: ii).

Regulation of clinical practice

The nature of the international collaborative effort to decipher the human genome would suggest the need for international agreement and approval of a common policy for use of the resultant data in the application of genetic engineering for the alleviation of disease. Such a policy would need to impose strict controls on access to data bases in order to safeguard against possible misuse of information, and it would be desirable to formulate a strategy whereby research findings could be made available for the mutual benefit of all nations. Prioritization of needs could also be decided on an international basis so that undeveloped countries were not excluded from the advantages to be gained. However, complex problems, previously identified, may prevent such equality of opportunity for all. It could be envisaged that new international law will become necessary to ensure that the possession and application of genetic information is not used as a means of compounding political or economic differences between nations.

On a national basis, a controlling body, with similar functions to the Human Fertilization and Embryology Authority, could exist to authorize the granting of licences, maintain records, oversee research and regulate the practice of modification of germ cells. A recent report (*Report of the Committee on the Ethics of Gene Therapy* 1992: 19–27) recommended, in anticipation of somatic gene therapy being practised in the near future, the establishment of a supervisory body 'of sufficient standing to command the confidence of local research ethics committees, and of the public, the professions and of Parliament' and proposed that it be non-statutory in order to retain flexibility. Similarly, in relation to such therapy, a joint statement by the Medical Research Councils of several European countries[1] advised that an expert national body should consider and approve all proposals, to ensure that agreed national guidelines are applied and public approval elicited.

Finally, the practice of genetic correction would raise a number of problems which would need to be addressed by law. Current legislation, particularly on consent to protect the interests of future generations and on access to information in order to prevent discrimination by employers or insurers on the basis of genetic credentials, may have to be amended to deal with the demands of increased litigation, involving 'breached confidentiality, genetic damage, wrongful life and medical negligence', which will be almost inevitable with new techniques and procedures distributed non-uniformly upon differently trained and experienced health workers (Friedmann 1990: 412).

CONCLUSION

Genetic alteration of the germ-line, such as that involved in engineering disease resistance, may become feasible in the future. Although there is a consensus of opinion that it should not be contemplated, it should not be dismissed totally on a basis of unjustified fears about its possible misuse or risks of disastrous irreversible changes. Decisions should only be reached after deliberation of the efficacy of the procedure after extensive research, assessment of possible benefits against risks, considerations of ethical issues which might arise and means of protecting against misuse by strict regulation of practice.

As techniques for genetic engineering become more advanced, so the possibilities of extending those techniques to the germ-line will be increasingly debated. Although methods for human experimentation are strictly controlled, the possibility of unfavourable consequences can never be eliminated, but such fears cannot necessarily provide sufficient justification to prevent a form of scientific progress which may be highly beneficial:

> great benefits have followed from people's willingness to consider the worst cases, and then to decide deliberately how they must be avoided. This is the spirit in which genetic manipulation has been, from the outset, approached. To follow a different course in this connection will have the effect of ensuring that those who may responsibly worry about what may emerge remain in ignorance, leaving the field open to those who are less responsible.[2]

Marcer (1991: 209) proposes the need for an ethic of long-range responsibility, implying a 'moral imperative to obtain predictive knowledge and data about the wide-ranging possibilities of some action', to future generations, that is, to beneficiaries and those at risk as yet non-existent. He suggests that genetic engineering allows a widening of our moral horizons beyond the traditional view of morality, which involves only the short-term consequences of our actions:

> persons in different generations have duties and obligations to one another just as contemporaries do. The present generation cannot do as it pleases but is bound by the principles that would be chosen in the original position to define justice between persons at different moments in time.
>
> (Rawls 1972: 293)

Respect for future generations and the opinions which they might hold must be a predominant concern in any debate on the moral permissibility of a revolutionary medical procedure which will directly affect both.

NOTES

1 'Gene therapy in man – recommendations of European medical research councils', *Lancet* (4 June 1988), 1272.
2 Editorial: 'Are germ-lines special?', *Nature* 331 (14 January 1988), 100.

REFERENCES

Anderson, W. F. (1985) 'Human gene therapy: scientific and ethical considerations', *J. Med. and Philosophy* 10.

—— (1989) 'Human gene therapy: why draw a line?', *J. Med. and Philosophy* 14.

Anderson, W. F. and Fletcher, J. C. (1980) 'Gene therapy in human beings: when is it ethical to begin?', *New Eng. J. Med.* 303 (27 November).

Budiansky, S. (1983) 'Churches against germ changes', *Nature* 303 (16 June).

Cooper, D. and Barefoot, M. (1987) 'Can you buy insurance for your genes?', *New Scientist* (16 July).

Dawson, K. and Singer, P. (1990) 'The human genome project: for better or worse?', *Med. J. Australia* 152 (7 May).

Friedmann, T. (1990) 'Opinion: the human genome project – some implications of extensive "reverse genetic" medicine', *Am. J. Hum. Genet.* 46.

Galloway, J. (1990) 'Britain and the human genome', *New Scientist* (28 July).

Glover, J. (1984) *What Sort of People Should There Be?*, Harmondsworth: Penguin.

Glover, J. *et al.* (1989) *Fertility and the Family – The Glover Report on Reproductive Technologies to the European Commission*, London: Fourth Estate.

Harris, J. (1992) *Wonderwoman and Superman: The Ethics of Human Biotechnology*, Oxford: Oxford University Press.

Joyce, C. (1990) 'Your genome in their hands', *New Scientist* (11 August).

Macer, D. (1991) 'Whose genome project?', *Bioethics* 5(3).

Motulsky, A. G. (1983) 'Impact of genetic manipulation on society and

medicine', *Science* 219 (14 January).
Rawls, J. (1972) *A Theory of Justice*, Oxford: Clarendon Press.
Report of the Committee on the Ethics of Gene Therapy (1992) (January) London: HMSO.
Roberts, L. (1990) 'Genome project: an experiment in sharing', *Science* 248 (25 May).
Watson, J. (1990) 'The human genome project: past, present and future', *Science* 248 (6 April).
Wertz, D. C., Fletcher, J. C. and Mulvihill, J. J. (1990) 'Medical geneticists confront ethical dilemmas: cross-cultural comparisons among 18 nations', *Am. J. Hum. Genet.* 46.
Yoxen, E. (1986) *Unnatural Selection? Coming to Terms with the New Genetics*, London: Heinemann.
Zohar, N. J. (1991) 'Prospects for "genetic therapy" – can a person benefit from being altered?', *Bioethics* 5:4.

8

HOW TO ASSESS THE CONSEQUENCES OF GENETIC ENGINEERING?

Heta Häyry

Assuming that there are no intrinsic ethical grounds for condemning biotechnology and genetic engineering,[1] the moral status of these practices can only be determined by an evaluation and assessment of their expected consequences. In theory, such an assessment should not be too difficult to carry out, since the advantages and disadvantages of gene-splicing techniques are fairly well publicized.[2] In practice, however, the situation is different. The actual consequences of applying advanced molecular biology depend upon the social and political setting in which the application takes place. As different groups of people hold different views concerning the structure and dynamics of social and political life, these groups inevitably also disagree upon the consequences of the development of recombinant DNA techniques.

My aim in this paper is to sort out and analyse, first, the major consequences of genetic engineering, and second, the most important ideological factors which cause dispute concerning their weight. I shall begin by reviewing the advantages and disadvantages of various forms of biotechnology as they have been presented by the proponents and opponents of these practices. I shall then go on to consider the different ways in which the effects of genetic engineering can be evaluated, depending on the degree of optimism or pessimism inherent in the referee's judgment.

THE ADVANTAGES OF GENETIC ENGINEERING

The advantages of biotechnology, as seen by its proponents, include many actual and potential contributions to medicine, pharmacy, agriculture, the food industry, and the preservation of our natural environment.[3]

Within medicine and health care, the most far-reaching consequences are supposed to follow from the mapping of the human genome, an enterprise which is still in its infancy but which holds the key to many further developments.[4] Accurate genetic knowledge is a precondition for many foreseeable improvements in diagnoses and therapies as well as an important factor in the prevention of hereditary diseases, and possibly in the general genetic improvement of humankind. Provided that such knowledge becomes available, potential parents can in the future be thoroughly screened for defective genes and, depending on the results, they can be advised against having their own genetic offspring, or they can be informed about the benefits of pre-natal diagnosis. One of these benefits is that an adequate diagnosis may indicate a simple monogenic disease in the foetus which can be cured by somatic cell therapy at any time during the individual's life. Another possibility is that an early diagnosis reveals a more complex disorder which can be cured by subjecting the embryo to germ-line gene therapy – this form of pre-natal treatment also means that the disorder will not be passed down to the offspring of the individual. Even in the case of incorrigible genetic defects the knowledge benefits the potential parents in that they can form an informed choice between selective abortion and deliberately bringing the defective child into existence.

Pre-natal check-ups and therapies do not by any means exhaust the medical applications of future genetic engineering. Somatic cell therapies are expected to help adult patients who suffer from monogenic hereditary diseases, and the increased risk of certain polygenic diseases can be counteracted by providing health education to those individuals who are in a high-risk bracket. In addition, the purely medical benefits of genetic engineering are extended by the fact that gene mapping will probably prove to be useful to employers and insurance companies as well as to individual citizens. Costly mistakes in employment and insurance policies could be avoided by carefully examining applicants' tendencies towards illness and premature death prior to making the final decisions.

Gene therapies and genetic counselling are practices which require advanced knowledge concerning the human gene structure. But the genetic manipulation of non-human organisms can also be employed to benefit humankind. The applications of gene splicing to pharmacy, for instance, may in the future produce new diagnosing methods, vaccines and drugs for diseases which have to date been incurable, such as cancer and AIDS. The agricultural uses of biotechnology include the development of plants which contain their own pesticides. In dairy farming, genetically engineered cows give more milk than ordinary ones, and with the right kind of manipulation the proteins of the milk can be made to agree with the digestive system of those suffering from lactose intolerance. As for other food products, biotechnology can be applied to manufacture substances like vanilla, cocoa, coconut oil, palm oil and sugar substitutes. And biotechnology can even provide an answer to the growing environmental problems: genetically engineered bacteria can be used to neutralize toxic chemicals and other kinds of industrial and urban waste.

THE DISADVANTAGES OF GENETIC ENGINEERING

The disadvantages of biotechnology, as seen by its opponents, are in many cases closely connected with the alleged benefits. An efficient strategy in opposing genetic engineering is to draw attention to the cost and risk factors which are attached to almost all inventions and developments in the field.

One general critique can be launched by noting that the applications of recombinant DNA techniques are enormously expensive. Millions of pounds are spent every year by governments and multinational corporations in biotechnological research and development. These resources, opponents of the techniques argue, would do more good to humankind if they were allocated, for instance, to international aid to the Third World.

Another problem is that, despite the undoubtedly good intentions of the scientists, the actual applications of genetic engineering are often positively dangerous. Consider the case of plants which are inherently resistant to diseases, or which contain their own pesticides. Although there are no theoretical obstacles to the production of such highly desirable entities, corporations – who also sell chemical pesticides – might prefer to market another type of genetically manipulated plant, which is unprotected against pests but highly tolerant to toxic chemicals. The result of this policy would be an increase in the use of dangerous chemicals in agriculture, particularly in the Third World – which is to say that the outcome is exactly opposite to the one predicted by the proponents of biotechnology. Besides, it is quite possible that genetically engineered grains are less nourishing than the grains which are presently grown. If this turns out to be the case, then the employment of biotechnology will intensify instead of alleviating famine in the Third World. And to top it all, genetically manipulated plants may contain carcinogenic agents, and thus contribute to the cancer rates of the developing countries.

An oft-used criticism against agricultural biotechnology is that the introduction of altered organisms into the natural environment can lead to ecological catastrophes. Scientists working in the field of applied biology have themselves noticed this danger, and set for themselves ethical guidelines which are designed, among other things, to minimize this particular risk. But as the opponents of genetic engineering have repeatedly pointed out, not all research teams follow ethical guidelines if the alternative is considerable financial profit.

Apart from the excessive expense and the increased risk of physical danger caused by biotechnology, the opponents of gene splicing can appeal to yet another disadvantage, which is related to widely shared moral ideals rather than to straightforward estimates concerning efficiency. Genetic engineering, it can be argued, is conducive to economic inequity and social injustice, both nationally and globally. Even in the most affluent western societies, gene therapies are too expensive to be extended to members of all classes and age groups. Subsequently, these therapies are likely to become the privilege of an elite, and they will drain scarce resources from the more basic areas of health-care provision. In the developing countries, the situation is even more absurd. Medical problems which originally stem from lack of democracy, lack of education, shortages of fresh water, population explosion, archaic land-ownership arrangements and the like cannot possibly be solved by high-tech western innovations which can barely be made to work in the most affluent and democratic of countries.

Another type of injustice emerges from the fact that the natural national products of many developing countries are superseded in the market by the biotechnological products of multinational corporations. Genetically engineered substitutes for sugar, for instance, could adversely affect the lives of nearly fifty million sugar workers in the Third World. Biotechnological vanilla could increase the unemployment figures by thousands in Madagascar, Reunion, the Comoro Islands and Indonesia. And plans to produce cocoa by genetically manipulating palm oil threaten the current export market of three poverty-stricken African countries, namely Ghana, Cameroon and the Ivory Coast. In all these cases, the profits of multinational Western corporations are clearly and directly drawn from the national income of the developing countries.

Finally, as regards medical biotechnology, there are those who believe that advanced knowledge concerning the human genome will inevitably become an instrument of genetic programming, which in its turn leads to subtle forms of genocide and general injustice. The opponents of gene splicing argue that the development begins inconspicuously with attempts to eliminate hereditary diseases. This practice of what is called 'negative' eugenics will, however, soon be followed by more 'positive' efforts towards altering the human genome: the inborn qualities of future individuals will first be improved in their own alleged interest, and then, later on, in the interest of society at large. When this development has gone far enough, scientists will also be asked to design special classes of sub-human beings who can do all those occupations which are too dangerous or too tedious for ordinary people. The outcome, according to the opponents of biotechnology, will be something like Aldous Huxley's Brave New World.[5]

TECHNOLOGICAL VOLUNTARISM V. TECHNOLOGICAL DETERMINISM

When one compares the views for and against biotechnology, one can easily see that the prevailing differences of opinion are not purely factual in nature. It is an undeniable fact that genetic engineering can be employed to eliminate diseases, but it is also an undeniable fact that genetic engineering can create bizarre and dangerous life forms. Likewise, it is possible that agricultural biotechnology will relieve famine in the Third World, but it is also possible that it will intensify it.

One ideological difference between those who defend recombinant DNA techniques and those who oppose them is based on their attitudes towards what have been called in the literature 'technological imperatives'.[6] The proponents of genetic engineering tacitly assume that the development and implementation of new techniques can always in the end be controlled and steered by human decision-making. If this were the case, then there would be no binding imperatives dictated to us by technological processes. This view has been labelled in the literature 'technological voluntarism'. On the other hand, however, the opponents of genetic engineering have argued that technological development in fact has its own internal laws and logic, which cannot be altered or checked by human choices or human action. This view, which can be called 'technological determinism', implies, among other things, that all new techniques, however dangerous or morally offensive they may be, will ultimately be implemented.[7]

Those who believe in technological determinism can support their views by referring to real-life examples drawn from the history of science and technology. The splitting of the atom, for instance, made it possible to devise nuclear explosives, and despite the expected evil consequences, the atom bomb was constructed and used as soon as the technical difficulties had been solved.[8] In medicine, the drastic improvement of life-sustaining therapies since the 1950s has led to a situation where irreversibly unconscious and intolerably painful human lives are prolonged beyond all reasonable limits, the only boundaries being set by medical technology. These and many other examples seem to show that every technological possibility will ultimately be exploited whatever the effects on human well-being.

Those who place their trust in technological voluntarism like to point out, however, that the 'imperatives' created by scientific know-how can always be rejected. Atom bombs were not launched by abstract technological systems, but by political decision-makers. Hospital patients are not kept alive against their own wishes and beyond any hope of recovery by medical technologies as such, but by doctors and medical teams whose actions are sanctioned by democratically chosen legislators. Even if it is true that technical solutions often 'offer themselves' to those making important decisions, these 'offers' can be ignored or rejected if the consequences of accepting them seem to be harmful or otherwise undesirable. The fact that technological imperatives can be disobeyed makes it possible, according to the voluntarist, for us to reject those forms of technology which are harmful while preserving those forms which are useful to humankind.

Both technological determinism and technological voluntarism appear to have some truth in them. There are cases in which humankind seems to be guided, to a certain extent at least, by scientific and technical inventions. But there are also cases in which scientific knowledge and its applications are more or less firmly in our control. To name only one example of each class, the atom bomb was built and used in the Second World War, whereas the development and use of poison gases was foregone by the political and military decision-makers of the period. These and many other real-life examples indicate that the dispute between determinism and voluntarism cannot be solved at a general level. But they also indicate that technologies can be divided into those which support the determinist interpretation and those which uphold the voluntarist view.

Given that technologies differ from each other in this respect, the important question in our present context concerns the category into which genetic engineering can be placed. Is genetic engineering one of those technological advances which control our actions, or does it belong to the class of techniques that we can control for our own benefit?

The empirical evidence which is presently available is mostly reassuring, and allows for moderate optimism concerning the controllability of biotechnology. Although there have been a number of unauthorized and potentially dangerous experiments with recombinant DNA, the guidelines set first by the scientists themselves and later on by governmental bodies have been rather leniently followed by the majority of research teams.[9] During the last two decades, gene splicing has been transformed from invention to industry, but up to this time human decisions have definitely played an important role in the development and application of the technique.

THE THREE STAGES OF TECHNOLOGICAL DEVELOPMENT

Granted that biotechnology has not yet taken total control over human affairs, the pessimist can nevertheless argue that the situation will in the near future be drastically different. The pessimist's view can be based on the idea that new techniques develop gradually, each stage provoking its own peculiar responses in the business world and among the general public.[10]

The first stage of technological development is the theoretical invention of a new technical tool or process. Depending on the nature of the invention, this stage can either go unobserved by all but a few experts in the field, or it can become widely publicized and commented upon. The latter is often the case when the new technique in question is connected, however remotely, to human sexuality and reproduction. At the second stage of technological development the newly invented tools and processes are transformed by practical innovations into industrial products and means of production. Since this phase usually extends over a relatively long period of time, public attention tends to fade away, along with the fears which may initially have been caused by the novel device. At the third and final stage the completed products are marketed and distributed among the consumers. Assuming that the preceding innovation phase has been long enough, the products often enter the market without protestation, regardless of the reactions which may have accompanied the introduction of the new technique.

The three-stage model of technological development can be employed by the opponents of genetic engineering to construct an argument against scientific and industrial gene splicing. When recombinant DNA techniques were first invented in the 1970s, their potential applications were publicly discussed and widely condemned. The reaction was for the most part based on religious taboos and on horror images created by twentieth-century science fiction, not on any real-life facts. The practical outcome of the protest was, however, that scientists between themselves decided to proceed cautiously, and to avoid arousing further popular outrage. Under the cover of professional ethical codes, biotechnology was then gradually advanced to its present innovational stage, at which the actual and potential industrial applications have become numerous and diverse. Attitudes towards gene splicing have, in the course of this development, inconspicuously become more permissive, largely because of the fact that there have been no major catastrophes.[11]

According to technological pessimists, the relaxation of popular attitudes marks the beginning of a new and increasingly dangerous era in genetic engineering. When scientists are no longer censoriously controlled by the general public, their readiness to follow voluntarily assumed guidelines will decrease, and their codes will be gradually slackened. At first the reasserted freedom of the scientists will be qualified by assessments of the expected harms and benefits. Although the majority of gene-splicing activities will be sanctioned, those forms of genetic engineering which are considered to be particularly hazardous will be kept under the closest surveillance. After a while, however, the qualifications will be removed. By the time biotechnological development fully reaches its third stage – the stage at which completed products are marketed and distributed –

consumers will have become accustomed to whatever forms of genetic engineering scientists care to pursue. And at that point, so the opponents of biotechnology claim, humankind will only be one step away from the Brave New World.

There are two reasons to doubt the validity of the pessimistic view. First, as regards empirical facts, most western countries will during the 1990s enter a phase where genetic engineering becomes regulated by law. The emergence of legal regulations undermines the immediate importance of public attitudes and voluntary guidelines, and the progress from qualified to unqualified licence predicted by the pessimist cannot take place. Second, even if the pessimist's prediction turned out to be correct, it remains unclear whether the unrestricted practice of gene splicing would in fact lead to the kind of genetic totalitarianism that Huxley described. One cannot reasonably assume that biotechnology could by itself destroy the prevailing democratic institutions and replace them by an authoritarian system. It is true that decision-making concerning the development of genetic engineering is at present controlled by multinational corporations rather than democratically elected governments. But it does not follow from this fact that humankind would be on the path to genocide and totalitarianism.

It seems to me that there are two types of pessimism involved in the issue, one of which is more readily justifiable than the other. The first, legitimate type of pessimism concerns the power of biotechnology to solve global problems like famine and pollution. Since it is not in the interest of multinational corporations to organize international aid to Third World countries, it is indeed unlikely that suffering would in the near future be effectively alleviated by implementing gene-splicing techniques. But this is not to say that genetic engineering ought to be banned, or that the situation could not be different sometime in the future. The second, illegitimate type of pessimism cynically assumes that nothing can be done to change the situation, and that the total prohibition of gene-splicing activities is the only way to save humankind from the slippery slope to which mad scientists and big corporations are leading us. This attitude is groundless and, moreover, positively dangerous in that the prophesies of the pessimist of this second type can easily become self-fulfilling. If ordinary citizens do not even attempt to control the corporation giants to make the benefits of biotechnology available to everybody, it is obvious that they will not attain such control. Pessimism in this sense is a form of ideological defeatism rather than considered intellectual scepticism.

WHAT ARE THE CONSEQUENCES?

To return to the consequences of genetic engineering and their assessment, it is now easy to see why competing views on biotechnology are so often so uncompromisingly antagonistic to each other. Factual disputes, which could be reconciled by empirical observation, play a relatively minor part in the controversy, and the real disagreement can be traced to commitments to optimism or pessimism, voluntarism or determinism. It seems, moreover, that these commitments are unavoidable, i.e. that it is impossible to produce a genuinely impartial assessment of the consequences of genetic engineering. It is therefore necessary to assess the validity of the underlying views before an evaluation of biotechnology can be completed.

As regards technological voluntarism and technological determinism, the empirical evidence which is available at present seems to indicate that genetic engineering is, by and large, controllable by human decision-making. The temptations created by technological possibilities should not be ignored or mitigated, but they should not be exaggerated either. Like all advanced technological innovations, gene splicing offers many possibilities, some of which will appeal to certain unscrupulous research groups despite the inherent hazards and despite the voluntary codes agreed upon by more responsible scientists. But the legal regulation of biotechnology, which in most western countries will take effect during the 1990s, provides an adequate device against such unethical conduct.

The issue of optimism and pessimism is slightly more complex, since different variations of both views can be either acceptable or unacceptable in the context of biotechnology. To begin with the gloomier views, epistemic pessimism is justifiable when it prevents people from believing uncritically what genetic engineers tell them about the advantages of the new techniques. Graver types of ideological pessimism cannot, however, be condoned, since they are prone to be self-fulfilling: widespread resignation and defeatism is likely to produce consequences which would never have occurred if people had been more hopeful. Similar observations can be made about the variants of optimism. It would be rather naive to believe that the multinational corporations which presently control biotechnological development would voluntarily undertake to further general welfare and global justice. Epistemic optimism in this sense is therefore unwarranted. But it is not necessarily naive to believe that changes are possible in economic and political life, and that genetic engineering can in the future serve wider interests than it does at the present time. Ideological optimism is in fact supported by the feature which can be used to disqualify its pessimistic counterpart: if the belief in a better world is in any way self-fulfilling, it is obviously better to hope and prosper than to despair and perish.

Assuming that democratic decision-making can be gradually extended to biotechnology, and assuming that genetic engineers continue to be controlled by voluntary agreements or legal regulations, the consequences of gene splicing can be expected to be prima facie good and desirable. Democracy both within and outside the world of biotechnology guarantees that no Brave New Worlds will be created as a result of misguided social experiments or malevolent dictatorial schemes. Surveillance and regulation, in their turn, provide a safeguard against reckless and inhumane research on areas such as biological warfare, the manipulation of our natural environment, and designs for new breeds or hybrids of animals. Granted that these possibilities are excluded, the remaining forms of genetic engineering are without doubt prima facie acceptable.

There are, however, certain embarrassing points which ought to be accounted for before recombinant DNA techniques can be wholeheartedly condoned. The moral problem at the core of the issue is that whatever advantages may ensue from the employment of genetic engineering, these advantages will not be allocated to those who need them most. The medical advances will almost exclusively benefit the wealthy and the privileged of the affluent western societies, and the industrial applications will mainly profit the big corporations, sometimes at the expense of destitute Third World countries. The prima facie desirable consequences of biotechnology are not genuinely worth pursuing, one might argue, unless they can be equitably distributed among nations and among individuals.

But although it is easy to point out instances of injustice in the application of gene splicing, none of these inequities is primarily caused by genetic engineering. The allocation of scarce resources is a growing problem in medicine and health care, but it has not been either originated or, to any significant degree, aggravated by the development of biotechnology. Gene therapies are not the only form of treatment which has been confined to the select few in our present-day societies. And although it is true that certain Third World countries will be even worse off than they are at present after multinational corporations have taken over their export markets, the roots of their suffering surely lie deeper. The invention of recombinant DNA techniques is, after all, a relatively minor evil compared to the centuries of western colonialism and imperialism which have left the national economies of these countries exceedingly vulnerable to any sudden changes in the world market.

What these remarks amount to is that genetic engineering should not be blamed for the injustice related to its implementation. If humankind can find a way to solve the problems of just distribution at a general level, then the benefits of biotechnology will also be distributed justly among individuals and nations. According to the moderately optimistic view that I have put forward and defended here, our best chance is to believe that this is possible, and then proceed to make it possible by our own actions. If this can be done, and if the requirements of democracy and adequate control are fulfilled, then the conclusion can be drawn that, in the last analysis, the expected consequences of genetic engineering favour its acceptance.

NOTES

I should like to thank Mark Shackleton, Lecturer in English, University of Helsinki, for improving the style of this paper.

1 See Matti Häyry's 'Categorical objections to genetic engineering: a critique', also in this volume.
2 For a balanced account of the advantages and disadvantages, see, e.g., P. Wheale and R. McNally, *Genetic Engineering: Catastrophe or Utopia?* (Hemel Hempstead: Wheatsheaf Books, 1988); P. Wheale and R. McNally (eds), *The Bio-Revolution: Cornucopia or Pandora's Box?* (London: Pluto Press, 1990).
3 See, e.g., Wheale and McNally, *Genetic Engineering*; Wheale and McNally (eds), *The Bio-Revolution*.
4 See, e.g., S. Kingman, 'Buried treasure in human genes', *New Scientist*, 8 July 1989; reprinted in *Bioethics News* 9 (1990): 10–15.
5 A. Huxley, *Brave New World* (Harlow, Essex: Longman, 1983 – first edition 1932).
6 I. Niiniluoto, 'Should technological imperatives be obeyed?', *International Studies in the Philosophy of Science* 4 (1990): 181–9.
7 Niiniluoto, op. cit., 182–4, makes an additional distinction between

pro- and anti-technological variants of technological determinism and voluntarism. In what follows I shall confine my comments to pessimistic anti-technological determinism and optimistic pro-technological voluntarism.

8 J. Ellul, *The Technological Society* (New York: Alfred A. Knopf, 1964).

9 The situation in the United States up to the end of the 1980s is described in D. M. Koenig, *The International Association of Penal Law: United States Report on Topic II: Modern Biomedical Techniques* (Lansing, Mich: The Thomas M. Cooley Law School, undated). The unauthorized experiments are reviewed by Koenig in note 122, on p. 30. Many European countries are now preparing legal regulations on gene splicing.

10 The following is mostly based on Niiniluoto, op. cit., 185, who is broadly citing the ideas of Joseph Schumpeter.

11 In the words of Daniel Callahan: 'It's very hard to sustain a great deal of worry about these things when, after ten years of pretty constant interest and attention, there have been no untoward events.' Cited by Koenig, op. cit., p. 29.

9

WHO OWNS MO?

Using historical entitlement theory to decide the ownership of human derived cell lines

Charles A. Erin

Advances in biotechnology over the past two decades or so have brought in their train novel legal and ethical problems. In particular, the use of human tissues and cells raises some of the most complex and sensitive issues, none more so in the commercial environment which the biotechnology industry inhabits than the questions of property and ownership.

Are human tissues and cells property? If so, who owns them? The answers to such questions clearly have far-reaching implications for patients and research subjects, medical practitioners and research scientists, not to mention philosophers and jurists, but I will not attempt to derive solutions here.[1] Rather, I will address a subordinate question which is crucial to the legitimacy of research involving human biological materials.

These materials may be used to produce cultured cell lines. A cell line is defined as:

> a sample of cells, having undergone the process of adaption to artificial laboratory cultivation, that is now capable of sustaining continuous, long-term growth in culture.[2]

Cell lines have wide-ranging applications in the biological sciences and may be of commercial value. Who owns, or ought to own, human-derived cell lines, and what constitutes a legitimate mode of acquisition of tissues and cells from a living person? I will take as an exemplar of the issues raised an actual case which has led to a major legal dispute.

BACKGROUND

In 1976, doctors at the University of California, Los Angeles (UCLA) School of Medicine removed the grossly enlarged spleen of Mr John Moore as part of his early treatment for hairy cell leukaemia (HCL), a rare form of cancer.[3] Following splenectomy, Moore's condition quickly improved.[4] Research by UCLA scientists David Golde (who was also the physician treating Moore) and Shirley Quan on cells of Moore's spleen showed his disease to be a T-cell variant of HCL,[5] very rare,[6] and they established a productive cell line, the 'Mo' cell line.[7]

Mo produces several substances of value to biological research,[8] and it was Mo which provided the supernatant from which scientists at the US National Cancer Institute isolated HTLV-II,[9] a new human T-cell lymphotropic virus which caused Moore's disease.

A patent was filed on Mo in 1981 and granted to the university in March 1984 naming Golde and Quan as inventors.[10] Meanwhile, Moore continued to return to Los Angeles to supply further blood samples,[11] apparently as part of routine follow-up procedure. When, in 1982, it became clear that Mo harboured HTLV-II, Golde asked Moore to become a subject for research on the virus, to which he agreed and gave consent.[12]

On at least one occasion Moore signed a form which gave the university rights to all cell lines, as confirmed by one of Moore's attorneys.[13] On another occasion, however, it seems Moore refused to grant the university rights in his blood and any cell line derived from it. Moore has stated that if Golde had not pointed out to him, in September 1983, that he had 'missigned' that consent form (by circling 'do not' on the form), he would never have known that Mo existed.[14] It was then that he consulted an attorney.

On 11 September 1984, Moore filed suit in the Los Angeles County Superior Court against the University of California, Golde and Quan, Genetics Institute Inc., and Sandoz Pharmaceuticals Corporation.[15] He claimed that the physicians treating him had patented a cell line of possible commercial value which had been derived from his cells without his explicit consent, and that they had misappropriated his 'bodily substances'[16] and in so doing had violated his rights of ownership.

The Charges

Moore's lawyers claimed that the consent he made was invalid on the ground that, since no mention was made in it of the possibility of commercial use, it was not truly informed.

However, there is more to the lawsuit than this, as became apparent at the first court hearing on 29 October 1984, which dealt with procedural matters.[17] The university maintained that the case should be dealt with in the federal courts. Moore's attorneys concentrated on the lack of informed consent and the charge of 'conversion', and claimed the case should be adjudicated by the California state court, and this is how the judge ruled.[18]

Conversion

In American civil law, conversion is the equivalent of theft in criminal law. According to Diana Brahams, conversion is 'a tort of strict liability' and is:

> an act intentionally done inconsistent with the owner's right, though the doer may not know of or intend to challenge the property or possession of the true owner.[19]

This translates to the layman as a form of misappropriation which is particularly unforgiving towards the misappropriator in that, unlike theft, it does not imply knowledge or intent on her part.

The Courts

There is very little by way of legal precedent applicable to the Moore case. A prior similar confrontation between Hideaki Hagiwara and the University of California, San Diego (UCSD) was settled out of court with the issue of ownership left in limbo.[20]

The Moore case has had a number of hearings, and various decisions have been handed down. I will briefly review the more significant of these.

The California Court of Appeals 1988

An earlier court ruling that Moore had no cause of action was overturned by the California Court of Appeals in July 1988 in a majority decision of two to one.[21] The court concentrated on the charge of conversion, which raised three questions: was the removed splenic tissue Moore's property?; did Golde take the tissue wrongfully?; and did Moore suffer damages?

Essentially, the court considered excised tissues to be the property of the individual from whom they are removed. The court saw no reason why, if physicians or scientists can own the tissues removed from an individual, the individual cannot; or why, if scientists and biotechnology companies can profit from those tissues, the individual should not share in those profits. It was ruled that Moore could bring his action.

It is evident that the court took a dim view of the biotechnology industry and of scientists' involvement with it:

> biotechnology is no longer a purely research oriented field in which the primary incentives are academic or for the betterment of humanity. Biological materials no longer pass freely to all scientists. As here, the rush to patent for exclusive use is rampant. The links being established between academics and industry to profitize [*sic*] biological specimens are a subject of great concern. If this science has become science for profit, then we fail to see any justification for excluding the patient from participation in those profits.[22]

There may be elements of truth in this opinion, but I will show below that the case against researchers, in the Moore case at least, is overstated.

The California Supreme Court 1990

In July 1990 the appeals court decision was subverted by the state of California's highest court.[23] The California Supreme Court handed down a final decision in regard of the demurrer.

The court decided not to entertain the concept that patients could own surgically removed tissues, and thus Moore's claim of conversion was disallowed. Several factors figured in this decision. For example, it was judged that to consider Moore's cells as property at common law would have been inconsistent with statute law, which indicated that a patient could have no more than a very limited right to control of excised cells. This conclusion has been criticized by Sir Eric Scowen:

> The statutory law in California exerted a strong influence on

> the Court, but in truth it is ambiguous in the present context . . . [and] . . . the statutory provisions do not necessarily deny Moore's conversion claim.[24]

As I will do below, the court distinguished between Moore's cells and the cell line, holding that, even if Moore could own cells removed at surgery, he could not own the line which was subsequently derived from them. Furthermore, it was considered that to grant property rights over excised cells would hinder the advance of research and that this was to go against the public interest.

Although the charge of conversion was thrown out, the court did decide that Moore had causes of action for issues related to breach of fiduciary duty and lack of informed consent. Moore claimed that Golde should have disclosed his research interests and any potential for commercial application before splenectomy. The court decided that these elements should be included in the scope of informed consent doctrine and the scope of a doctor's fiduciary duties. Moore may still lodge an appeal against the Supreme Court's ruling on the property element, but in the meantime he is suing over the fact that Golde did not inform him of the possible commercial use of his cells.

THE ISSUES

One of Moore's attorneys has been quoted as saying:

> the central issue in this case is the patient's right to a share of the profits earned by drug companies and biogenetic firms from products derived from his body.[25]

Although his opinion that 'the market potential for such products is believed to be in the billions of dollars'[26] as yet remains a vast exaggeration in the Moore case, the pecuniary aspect is an issue of some significance, no doubt. However, I think the first question we should try to answer is: who *owns* cell lines derived from human subjects? The alleged (legal) *right* of patients to share in potential profits seems to hinge on this.[27]

Three candidates with plausible claims to Mo may be identified immediately: (i) the individual who supplies the original material used to establish the cell line; (ii) the research workers who mix their intellectual ability and labour with this material to produce the cell line; and (iii) society generally, which stands to gain in terms of the advancement of science and enlargement of the biomedical knowledge base, and any consequent progress in therapeutics.

Two Approaches

Whether Moore owns his excised spleen tissue is clearly a critical factor in determining who owns Mo. However, there is no need to produce here an argument for the ownership of excised human tissue. I will argue that a cell line is neither a part nor a product of the human body. It will then be sufficient, for the purposes of deducing who owns Mo, to proceed from each of the two starting premisses: (i) John Moore owns his excised spleen tissue; and (ii) the excised spleen tissue is not his property. I will show that in this particular case the same conclusion is reached by either approach.

Cell Lines Are Not Products Of The Human Body

In the Hagiwara case, UCSD held that the cell line, CLNH-5, being a hybrid and not previously existing in nature, represented a new biological entity and was sufficiently dissimilar to its source to defeat any claim by the donor.[28] In the Moore case, Golde takes a similar tack and focuses on the essential nature of a cell line:

> A cell line is characterized by the single feature of immortality, which is *not* a property of normal human cells. Thus, by virtue of immortalization, the cell line is essentially different from the original tissue from which it is derived.[29]

There are usually many further differences between the derived line and the source cells, and the fact that patents were granted for both CLNH-5 and Mo supports this view of their novelty. A patent was not issued on Moore's splenic tissue but on what Golde *et al.* invented using the tissue as just one component.

Deriving A Cell Line

Although Moore's tissue was a vital ingredient in the mix, it was not the only vital ingredient. First, there is the 'hardware' to consider: culture media, laboratory equipment, etc. Access to specialist laboratory equipment is essential, particularly in view of safety requirements – although not known at the time of its derivation, Mo was later shown to harbour HTLV-II[30] and thus to represent a biohazard.

There is another obvious item without which Mo could not have been brought into existence: technical knowledge of a high level is a prerequisite for the establishment of cell lines.

However, we should not assume that the successful establishment of a cell line is merely a consequence of possessing the necessary technical know-how and having access to the appropriate equipment. Whilst there are many thousands of cell lines in existence today, there is an element of serendipity involved in their derivation. As Golde puts it:

> the process of deriving a cell line is poorly understood and even to this day is largely an art form involving the general skills of tissue culture. There are many elements involved in the immortalization and transformation of cells which can be accomplished by viruses and the uses of primary tumor tissue. In the case of the Mo cell line, while we think the cell line was skillfully derived, we now know that its derivability likely related to the fact that the patient harbored the HTLV-II virus, which was unknown until some five years after derivation of the cell line.[31]

Expertise of a high level is also required for a cell line's complete characterization,[32] again essential from a safety standpoint.[33] Continued research on Mo by the UCLA team[34] further demonstrates the high degree of specialist skill and esoteric knowledge required to make anything worthwhile of the cultured cells.

But this is not the full extent of the mixing of the researchers' intellectual ability with the cells. The initial decision to derive a cell line must be seen as the result of intellectual inquisitiveness and inventiveness over and above technical proficiency. It was the scientists' understanding of HCL and their recognition that the T-cell phenotype of Moore's hairy cells presented something new, combined with ingenuity on their part, that were the key factors in bringing Mo into existence. Moore's cells provided the spur which stimulated Golde *et al.*'s intellectual curiosity, but before they went to work on the cells, Mo did not exist, and if they had not looked at the cells, Mo would not have been brought into existence.

Having established and characterized the cell line, the researchers were then in a position to exploit this tool for the sake of medical science and expansion of the world knowledge base. This sounds very altruistic. I do not mean to portray the motives of the scientists in a Mertonian light[35] – to do so would be naive in the current environment of widespread commercialization of biotechnology. But the above conclusions are independent of the motives of the researchers, and the general benefit to humankind will be served whether they foresaw the possibility of financial return before engendering the cell line or whether this contingency presented itself after further research.

Notwithstanding Mo's derivation from cells of Moore's body, consideration of what a cell line actually is, and what is involved in its creation, leads me to conclude that a cultured cell line is not a product of the body. A cell line is not something that is pre-existent in nature. It is, rather, a human-made invention, an artefact. Moore's cells were, clearly, crucial to the derivation of Mo, but it was the researchers' idea to create Mo, and Moore had nothing to do with that idea, or with the work that went into bringing Mo into existence. Thus he cannot be considered in any way to be the creator or inventor of Mo.

ROUTE I: MOORE OWNS HIS EXCISED SPLEEN TISSUE

The Mo cell line cannot be considered a product of the human body. It was created, or 'begotten', by Golde and Quan from Moore's spleen tissue and various other items. Is this sufficient to invest the creators of Mo with title to it?

Where property originates – how titles originally arise – remains a philosophical puzzle. This is not the place to go into the minutiae of this vexing problem, but one theory which goes some way to answering the question is Historical Entitlement Theory (HET for short). Essentially, HET provides a means of deciding who holds title to things, by tracing ownership back to first occupancy.

Here I will adopt Hillel Steiner's formulation of HET's rules for generating titles.[36] According to HET's begetting rule, for the begetting of Mo to invest title to Mo with the begetters, they must own *all* of the items which went into the creation of Mo. The only item whose ownership is in question is Moore's tissue, ownership of which we are assuming lay originally with Moore. The important question here is: how did the researchers come into possession of the raw tissue which they used to engender the cell line? There are at least three modes of acquisition open to them.

Commercial Alienation

First, assuming that Moore has full rights of ownership including that of commercial alienation, he could have sold his excised spleen to the university. This did not happen. There are examples of patients selling their tissues,[37] but sale at the point of treatment would not have been a likely option for Moore.

Abandonment

Second, the researchers could have obtained Moore's spleen tissue through his abandonment of it. In law, abandonment in this sense is:

> The relinquishment of an interest or claim or of possession of property with intent to terminate proprietary interests therein, as by throwing away, or leaving and not seeking to retrieve it.[38]

At first glance, this may appear to be what happened in the Moore case. Most patients probably do not think of what is removed at surgery as being of potential value. By dint of the fact that it is medically indicated that my spleen needs to be removed, i.e. there is something wrong with it, it is malignant or whatever, I would not expect my interest in it to continue after removal. However, if I gave any thought to the destiny of my excised spleen, I would probably expect it to be incinerated.[39]

It would be uncharitable to attribute to John Moore the conscious act of abandoning his spleen. Indeed, even if you throw something away, or lose it, this does not necessarily count as abandonment: Diana Brahams refers to 'convictions for the theft of golf balls lost on club premises and of refuse from a dustbin'.[40]

Donation

Third, Moore could, of his own volition, have transferred title to his excised tissue to the university. It is abundantly clear from the literature that this is what happened. Moore consented to the use of his excised spleen for research purposes and he voluntarily gave over title to the removed tissue to the university.

It is thus HET's transfer rule which is relevant here. Simplistically, we can say that, by the signing of a consent form,

Moore transferred the title we have granted him in his spleen tissue to the research workers, and this is how they acquired title to it.

That Moore transferred title to his cells in this way is not in question: he admits this. But he says that he should have been informed of the possibility that his cells would be used to make something of potential commercial application before consent, before he divested himself of title to his cells.[41] This raises the question of *informed* consent, and in particular where the boundaries of disclosure should be drawn for consent to be valid.

Informed Consent

The doctrine of informed consent is underpinned by the principle of autonomy and serves to safeguard the well-being and right to self-determination of the patient or research subject. The aim is that, prior to any proposed therapy or trial which may present any reasonable possibility of risk, the patient or research subject, on the basis of information supplied, be put in a position such that she can make a maximally autonomous choice whether to proceed.

There are two dimensions to the consent Moore gave. First, there was the consent he gave for the splenectomy to be performed. Informed consent requires disclosure of the nature, probable consequences and possible dangers of, and alternatives to a proposed procedure, in clear terms, so that the patient is able to make a free, understanding decision whether to proceed with it. With no evidence to the contrary, I think we have to assume that this disclosure was made.

Second, there was the consent he gave for the research use of his excised tissue. It is not clear from the literature whether this was an extended consent or a special consent. Does it matter? What if Moore only gave consent to the splenectomy? Holder and Levine, writing prior to the Moore case, found that:

> In general, in obtaining consent for surgery for customary medical purposes, no special consent seems necessary to retain removed tissues or organs for research purposes. Of course, if the surgery is done for research purposes, this must be made clear in the consent discussion; however, in this situation the special consent is required to do the procedure rather than to retain the specimen.[42]

Why Did Moore Give Consent?

Moore did sign over title to his excised tissue to UCLA. Why? The motivation for such consent is usually taken as altruism: the hope on the part of the donor that research may realize any potential for benefiting humankind in some way. If this was Moore's hope when he originally gave his consent (before he had thoughts of pecuniary gain), then it has been realized. For example, a virus has been described which is the cause of a very rare form of leukaemia.

I have found no mention of Moore objecting to research using his cells or the cell line. Indeed, it is unlikely that Moore would have filed his suit if Golde *et al.* had not demonstrated their usefulness. The value (in whatever terms) of the cells is contingent on research being performed on them: whilst Moore's cells have instrumental value, they have no intrinsic value.

Profit Should Not Figure In Informed Consent Doctrine

Whilst the extent of disclosure may be a matter of controversy, even proponents of plenary disclosure would be hard pressed to justify disclosure of each and every particular research use to which, say, a blood sample might be put.[43] In many cases, the particular research use cannot be known until after investigation of the tissue.

Disclosure of a possibility of profitable product development (which possibility it is often impossible to predetermine) is irrelevant to the objects of informed consent. Thus I see no moral compulsion to divulge the potential for commercial spin-off, at least *not as an element of informed consent doctrine.* [44]

Conclusion I

In this case it seems that there was nothing perverse or fallacious about the consent obtained. Moore's transfer to the university of title to his cells was valid. By HET's transfer rule, the university acquired title to Moore's spleen tissue legitimately. Therefore, according to HET's begetting rule, because the university owned all of the ingredients which its agents, Golde *et al.*, used to create the cell line, the university owns Mo.

ROUTE II: MOORE DOES NOT OWN THE EXCISED TISSUE

Let us now follow the argument from the alternative starting premiss, that is, Moore does not own his excised spleen tissue. If Moore does not own the tissue, it seems fair to assume that at the time of removal it is unowned.

Applying HET's begetting rule (to the engendrure of the cell line), we have to show that Mo was created from items all of which were owned by the begetters. Again the critical question is: how did the researchers come by Moore's cells? I think that it is fairly straightforward to say they appropriated the cells. In terms of HET's appropriation rule, the researchers were the first to acquire ownership of something which previously was not owned. They did this by mixing their labour with the cells – not just their labour, as we have seen: it also involved the mixing of their inventiveness, intelligence, etc.

Conclusion II

UCLA's agents, Golde *et al.*, were the first acquirers of Moore's cells and so, by HET's appropriation rule, the university owns the cells; they then created a cell line from this and other ingredients, all of which the university owned; *ergo*, by HET's begetting rule, the university owns Mo.

Problems

But perhaps the argument proceeds too swiftly. There are two obvious objections to be raised here. First, HET's appropriation rule actually requires that the appropriator is the first to acquire the unowned thing *from nature*. Form Locke's *Second Treatise*:

> Whatsoever . . . he removes *out of the State that Nature hath provided, and left it in*, he hath mixed his Labour with, and joyned to it something that is his own, and thereby makes it his Property. [45]

Can we simply say that Moore's cells, once removed, reside in the state that nature has left them in? I will say that we can. Not just because it simplifies the argument, but because it seems to me that the main object of HET's appropriation rule is to provide a mode for the acquisition of unowned things, and the university's acquisition of Moore's unowned cells is certainly within the spirit of the appropriation rule.

The second problem stems from the appropriation rule's distributive proviso. From the same passage in Locke:

> this labour being the unquestionable Property of the Labourer, no Man but he can have a right to what that is once joyned to, at least where there is *enough, and as good left in common for others.* [46]

How should one interpret this proviso? Robert Nozick's interpretation goes like this:[47] All agents are invested with the liberty to use natural resources; if we assign equal liberty a certain value for each person holding it, then title to any natural resource lies with the agent who acquires it only if no other person is deprived of any of that value: one's appropriation of some thing should not worsen the position of anyone else by constricting their liberty to use that thing. On Nozick's interpretation, the upshot of this is that UCLA, as appropriator, is liable to a Lockean proviso-based tax of some sort: it owes compensation to all those who are subject to a loss of liberty-value.

I will not rehearse his argument here, but Hillel Steiner has shown that there are logical difficulties inherent in this interpretation.[48] Nozick's is just one of several interpretations of the proviso. Some writers would deny that there is a serious proviso in Locke at all. I will, in this instance, ally myself with this camp and say that it does not matter.[49] The university acquired the previously unowned cells legitimately and is thus the owner of Mo.

THE COMMONWEAL

What of the third candidate's claim to Mo, the general claim of humanity? Assuming that research on the cell line was to lead to advances in the understanding of HTLV-II or to a product which could, for example, retard the spread of the virus, this would be seen as a significant benefit to humankind, and in particular to those who suffer T-cell HCL. Could we then argue that there exists a general interest in Mo which is strong enough to support denying UCLA ownership of the cell line and instituting it as some form of public property? I think not. One must question what purpose this would serve. After all, the value of the cell line is not intrinsic, but lies in its status as a useful tool for biomedical research. The general benefit lies in successful research using the line and not in Mo *per se*.

Nevertheless, society's claim is worth considering for two reasons. First, if we take the general claim not as one to property in Mo but to the benefits which research on the line may yield, I think we are led to conclude that within the current legal context the commonweal is better served if rights of ownership of Mo remain with UCLA.

Limited Property Rights I

Second, we may wish to modify the researcher's rights of ownership when we take into account that society may stand to benefit in such a real sense as the control of a human retrovirus. For example, even though Mo is the private property of UCLA, we may wish to impose the restriction that rights of ownership may not extend to wilful destruction of the line.

Though temporary secrecy to protect ongoing research may be tolerated,[50] the utility of Mo would be maximized if the results of research performed using the line were freely disseminated worldwide for other interested scientists' use. Whilst speedy publishing of results in scientific journals is a chief goal of the scientific ethos, the cynical may wish to safeguard the general interest by ratifying a form of *glasnost* on research use of Mo as a qualification of its ownership.

The supply of samples of the cell line to other researchers in the field would provide for the greatest possible exploitation of this research tool and thus we might like to impose a requirement that the cell line be made available to other interested parties. Despite the belief of the California Court of Appeals quoted earlier, this already takes place, and Mo has been supplied *gratis* to other research establishments. There are necessary safety requirements: the availability of a potentially hazardous biological material needs to be regulated so that only those laboratories properly equipped to deal safely with it be given access. However, other limitations are introduced in practice. For instance, the 'mother' institution which holds the patent will usually require legal undertakings from recipient institutions that title remains with the supplier and that the material is not to be used for commercial purposes.

A Market in Human Organs and Tissues?

As of the California Supreme Court's ruling in the Moore case, it seems that my second starting premiss, that patients do not own tissues removed from their bodies, has been ratified as the current legal position, albeit by way of confirmation of the traditional legal tenor. It seems likely that *Mo* will be found wanting as a legal precedent, and it may be that at some future date an individual's ownership of her excised tissues will be ratified legally. The implied commodification of excised human tissue sails against the wind of legal tradition regarding the sale of human body parts and is not an immediate prospect.

However, if western legal tradition were to be overcome, and my first starting premiss – that individuals own their excised tissues – were to prevail, it seems to be an immediate consequence that some form of legitimate market in excised human tissues should be established. And it seems to be only equitable that patients and research subjects be given the opportunity to profit from the sale of tissues removed from their bodies. It is somewhat ironic that when those who object to the commercialization of human tissue procurement say that profiting from human tissues should not be allowed, what they mean is that individuals who supply tissues should not be allowed the option to profit from them – others involved in tissue procurement are in a position to profit.

Several market systems may be postulated.[51] I will not explore the various possible schemes here, but it should not be forgotten that a market does not necessarily rule out donation as a concomitant mode of tissue procurement.[52] Nor does it necessarily undermine the residual altruism involved in an individual's sale of her tissues.

Several problems have been predicted as inherent in a commercial system. For example, it is thought that just the fact of having such a system may induce patients to believe that their tissues are innately valuable, and this could lead them to make profit-motivated decisions incompatible with their health care. There are already fears that just the inclusion in the consent form of a clause mentioning possible research use could lead patients to view their bodies as money-spinners.[53] However, as the Moore case attests, there are problems with a non-commercial system, and it is certainly a possibility that a market could be established which functioned ethically and equitably at a practical level.

Limited Property Rights II

Accounting for the general claim of society led me to suggest that the researcher's absolute ownership of the cell line may need to be tempered. Does society's claim have any similar implications for the individual's supposed rights of ownership? John Harris, writing in a different context, states:

> deadly and infectious viruses and so on may grow on or in someone's body but it is not clear that they are owned in any sense that would preclude society's right to kill or otherwise dispose of them against the will of the 'owner' where the social utility of their destruction is clear.[54]

Conversely, might we not say that, where a clear social benefit lies in the preservation of such viruses, a cell line may be derived from the infected tissue, irrespective of the individual's property rights? I am not suggesting that researchers be empowered to excise tissues which they believe may be of research interest, but that, where the tissue has been removed as part of therapy (and, by implication, that tissue can serve the individual no good) and there is some likelihood that research on it may lead to medical advance, the individual's rights of ownership may not extend to refusal to transfer title. In effect, the individual may be forced to transfer title to his cells, whether by donation or by sale, to those who can reap some general benefit from them.[55]

Route I Revisited: John Moore Has Been Wronged

Applying HET's rules for generating title, I hope to have shown that, in this particular case and against the current legal background, even if Moore had owned his excised spleen, the university acquired title to the Mo cell line legitimately. However, if one accepts my first starting premiss, following through with its logical implications, it seems that individuals ought to be afforded the opportunity to sell their excised tissues. Then, at the moral level there is a case for saying that John Moore has been wronged, though not harmed,[56] by having been denied the opportunity to profit from his excised spleen. This is not to say that Moore has been wronged by Golde or UCLA, who, I hope to have shown, acted legitimately and without malice within the existing legal framework. John Moore has been wronged by the traditional legal attitude towards the individual's property rights in her living body and its parts and products.

NOTES

1 This essay is an abridged version of a part of an ongoing project which addresses these questions specifically and explores the implications of the answers to them.

2 Congress of the United States, Office of Technology Assessment, *New Developments in Biotechnology: Ownership of Human Tissues and Cells* – Special Report – OTA–BA–337 (Washington, D.C.: US Government Printing Office, 1987), p. 3, n. 1.

3 See, e.g., G. J. Annas, 'Whose waste is it anyway? The case of John Moore', *Hastings Center Report* (October/November 1988), 37.

4 In a faxed letter to *The Times* (London) (unpublished) dated 1 August 1990 Golde asserts:

> One fact is indisputable: removing Mr Moore's spleen saved his life!

As far as I am aware, this is indeed indisputable, but I do not see that it has any bearing on the issues raised by the case.

5 A. Saxon, R. H. Stevens and D. W. Golde, 'T-lymphocyte variant of hairy-cell leukemia', *Ann. Intern. Med.* 88 (1978), 323–6.

6 In fact, HCL due to proliferation of cells of T-lymphocyte origin had not been described, although, with the advantage of hindsight, Saxon *et al.* (ibid., 325) judge that a previously reported case of HCL (S. H. Advani, G. V. Talwalkar, J. S. Nadkarni, J. J. Nadkarni, S. M. Sirsat and U. Srinivasin, 'Hairy cell leukaemia, a case report and review of the literature', *Indian J. Cancer* 13 (1976), 283–7) may have been due to the proliferation of T cells.

7 Saxon *et al.*, op. cit., 325. For the researchers' method of establishing a continuous cell line see D. W. Golde, R. H. Stevens, Shirley G. Quan and A. Saxon, 'Immunoglobin synthesis in hairy cell leukaemia', *Br. J. Haematology* 35 (1977), 360.

It should be understood that there are two Mo cell lines, designated Mo-T and Mo-B. Mo-T is a line of immortalized T cells established from Moore's spleen tissue. The Mo-B line was established from peripheral blood cells about two years after the splenectomy, and is an Epstein–Barr virus-transformed B-lymphoblast cell line.

Mo-B was available to other interested researchers before Mo-T because the university wanted to wait for the patent on Mo-T to be granted before making it generally available (D. W. Golde, 'Availability of Mo and HTLV-II', letter, *Nature* 308 (1984), 19). It thus seems fair to infer two things: first, the patent on Mo refers exclusively to Mo-T; and second, the T-cell line is of greater interest in research terms. Where I refer to Mo, I mean Mo-T.

8 For example, immune interferon (type II), T-cell growth factor (interleukin-2), and GM-CSF (Granulocyte-macrophage colony-stimulating factor) (see, e.g., Barbara J. Culliton, 'Mo cell case has its

first court hearing' *Science* 226 (1984), 813).

9 V. S. Kalyanaram, M. G. Sarngadharan, Marjorie Robert-Guroff, I. Miyoshi, D. Blayney, D. Golde and R. C. Gallo, 'A new type of human T-cell leukemia virus (HTLV-II) associated with a T-cell variant of hairy cell leukemia', *Science* 218 (1982), 571–3.

Mo-T and Mo-B both harbour HTLV-II. In 1985 another case of atypical HCL, patient NRA, provided a second isolate of HTLV-II as reported by Golde's research team (J. D. Rosenblatt, D. W. Golde, W. Wachsman, J. V. Giorgi, G. M. Schmidt, J. C. Gasson and I. S. Y. Chen, 'A second isolate of HTLV-II associated with atypical hairy-cell leukemia', *New England Journal of Medicine* 315 (1986), 372–7).

10 Culliton, op. cit. at n. 8, 813.

11 Allegedly, it was only when UCLA offered to defray his expenses for these trips that Moore began to suspect that his blood was being used for more than just his personal treatment. Or at least that is the story we get from early reports of the case: see, e.g., Lori B. Andrews, 'My body, my property', *Hastings Center Report* 16 (1986), 28, where a newspaper report (Anon., 'Whose body is it anyway?' *Chicago Tribune*, 9 May 1985, p. 1A) is cited as the source of this account; but see also n. 12 below for a source of the alternative account of how and when Moore's 'suspicions' were aroused.

12 Moore later alleged that he did not know what sort of research was involved. Apparently, the fact that the consent he gave involved the signing over of his rights in any cell line derived from his cells led Moore to suspect that Golde was 'turning a profit from his cells' (Marcia Barinaga, 'A muted victory for the biotech industry', *Science* 249 (1990), 239).

13 Barbara J. Culliton, 'Patient sues UCLA over patent on cell line', *Science* 225 (1984), 1458.

14 Judith Stone, 'Cells for sale', *Discover* (August 1988), 34–5.

15 Why were a biotechnology company and a pharmaceutical company cited in Moore's suit? There is some controversy here. According to an article in *The Times* (Susan Ellicott, 'US court rules patients have no rights over removed body tissue', *The Times* (London), 11 July 1990, p. 9), Golde *sold* Mo for $3 million to 'a biotechnology company'. This was repeated by Diana Brahams in *The Lancet* (Diana Brahams, 'Ownership of a spleen', *Lancet* 336 (1990), 239). However, according to other reports (e.g. R. Eisner, 'Tissues not for sale', *Nature* 346 (1990), 208; Barinaga, op. cit. at n. 12), the university granted *licences* on Mo to Genetics Institute and Sandoz. Apparently Genetics Institute wanted to use Mo for cloning colony-stimulating factor (CSF), which is a lymphokine with potential commercial application as an 'immune-system booster', and they licensed the production of recombinant CSF to Sandoz. Golde became a member of Genetic Institute's scientific advisory board, for which he received 75,000 shares in the company (estimated to be worth $3 million). Golde says this was not in payment for Mo, but rather in compensation for seven years' scientific consultancy (D. W. Golde, 'Cell line ownership', letter, *Nature* 347 (1990), 419). The university received research grants from both companies worth a

total of £440,000.

Golde insists that he did not sell Mo, and since it was not his property and he did not hold the patent I do not see that he could have done. He has stated categorically (Golde, op. cit. at n. 4) that Mo has not been sold, that the university owns the patent, and that no licences have been issued.

16 Culliton, op. cit. at n. 13.

17 Moore's suit alleged several causes of action (see, e.g., R. Delgado and Helen Leskovac, 'Informed consent in human experimentation: bridging the gap between ethical thought and current practice', UCLA *Law Review* 34 (1986), 85, n. 79) including: conversion; lack of informed consent; deceit; breach of fiduciary duty; fraud; unjust enrichment; quasi-contract; bad faith; intentional infliction of emotional distress; negligent misrepresentation; intentional interference with prospective advantageous economic relationship; and slander of title.

18 Culliton, op. cit. at n. 8, 814.

19 Diana Brahams, 'A disputed spleen', *Lancet* ii (1988), 1151.

20 M. Sun, 'Scientists settle cell line dispute', *Science* 220 (1983), 393–4; I. Royston, 'Cell lines from human patients: Who owns them? A case report', *Clin. Res.* 33 (1985), 442–3.

21 The report of the court hearing is: *Moore* v *Regents of the U of California et al.*, 88 *Dailey Journal D A R* 9520 (Cal Cy App, 2d Dist, Div 4, 1988). However, I have not seen this and rely upon the account given by George Annas (Annas, op. cit., 38–9).

22 As quoted in Annas, op. cit., 38.

23 Eisner, op. cit. at n. 15; Barinaga, op. cit. at n. 12; Brahams, op. cit. at n. 15.

24 E. Scowen, 'The human body – whose property and whose profit?', *Dispatches* (Newsletter of the Centre for Medical Law and Ethics, King's College London) 1 (1990), 2.

25 Culliton, op. cit. at n. 8, 813.

26 Culliton, op. cit. at n. 13.

27 However, even if Moore does not (legally) own or have part ownership of the cell line, he may still have some moral claim to it independent of this.

28 Royson, op. cit. at n. 20, 442.

29 D. W. Golde, personal communication, 7 August 1989.

30 I. S. Y. Chen, Shirley G. Quan and D. W. Golde, 'Human T-cell leukemia virus type II transforms normal human lymphocytes', *Proc. Nat'l Acad. Sci. (U.S.A.)* 80 (1983), 7006.

31 Golde, op. cit. at n. 29.

32 E.g. A. Saxon, R. H. Stevens, Shirley G. Quan and D. W. Golde, 'Immunologic characterization of hairy cell leukemias in continuous cultures', *J. Immunol.* 120 (1978), 777–82.

33 From Golde, op. cit. at n. 6, 20:

> there can be considerable danger in providing cell lines before they are completely characterized.

34 E.g. I. S. Y. Chen, Jami McLaughin, Judith C. Gasson, S. C. Clark and

D. W. Golde, 'Molecular characterization of genome of a novel human T-cell leukaemia virus', *Nature* 305 (1983), 502–5.

35 R. K. Merton's idealistic characterization of the ethos of science is to be found in his celebrated paper 'Science and technology in a democratic order', *J. Legal Political Sociology* 1 (1942).

36 I am greatly indebted to Hillel Steiner for providing me with and discussing his account of HET's rules for generating titles. According to Steiner, HET imposes four rules for generating titles (H. Steiner, 'The fruits of body-builders' labour', pp. 64–78 above):

1 by *transfer* – by others transferring to us the ownership of things . . . they own, whether by gift or sale . . .;
2 by *appropriation* – by our being the first to acquire from nature some unowned thing (subject to certain distributive provisos . . .);
3 by *rectification* – by our acquiring ownership of things held by others who have violated our rights and consequently owe us those things in compensation;
4 by . . . '*begetting*' – by our making . . . something from/with things we already own.

37 For instance, Ted Slavin sold his blood, which was of interest to researchers and commercial enterprises alike because it had very high titres of antibodies to hepatitis B, for as much as $10 per millilitre.

It is significant that Slavin's marketing of his blood did not undermine his altruism: he also provided free samples of his serum to scientists at the Fox Chase Cancer Center. See B. S. Blumberg, I. Millman, W. T. London and others, 'Ted Slavin's blood and the development of HBV vaccine', letter, *NEJM* 312 (1985), 189.

38 D. M. Walker, *The Oxford Companion to Law* (Oxford: Clarendon Press, 1980).

39 Unless, of course, I had given my consent for its research use.

40 Brahams, op. cit. at n. 19, 1152.

41 That the Mo cell line will at some point realize its mooted *commercial* potential is, in fact, far from certain. It must be remembered that at the time UCLA applied for a patent on Mo there was much commercial interest in the production of interferons. Moore's spleen tissue provided an unusually abundant source of lymphokines – interferon is a lymphokine – and it may be that this was one of the factors which prompted UCLA to apply for a patent on Mo. The fact that a patent on Mo was obtained presumably reinforced conjecture as to potential profitability.

To date it appears that Mo's value lies solely in its research utility. It was reported in September 1984 that there were by then nearly one hundred other cell lines which were also prolific sources of lymphokines (Sandra Blakeslee, 'Patient sues UCLA over patent on cell line', *Science* 225 (1984), 1458), and more recently it has been reported that as yet neither Genetics Institute nor Sandoz have made any profits from selling Mo (Eisner, op. cit. at n. 15) – of course, if, as Golde maintains, no licences have been granted, they are not in a position to (see n. 15).

In this context the cell line is a tool. But the value of such a tool is not intrinsic. Rather, it is made valuable by the researchers' exploitation of it in a particular circumstance.

42 Angela Holder and R. J. Levine, 'Informed consent for research on specimens obtained at autopsy or surgery: a case-study in the overprotection of human subjects', *Clin. Res.* 24 (1976), 73.

43 There are cases where I would say that informed consent doctrine dictates that the specific research use to which biological material made available at surgery is to be put should be disclosed, such as in the consent of a woman whose abortus is to provide foetal neural tissue for transplantation. But these are special cases which raise different issues, and in general such considerations are not concomitant with protection of a donor's rights.

44 This holds only as long as there are no conflicts of interest on the part of the physician and the patient's treatment is not prejudiced and her rights of self-determination are not compromised by such considerations.

45 J. Locke, *Two Treatises of Government* (Cambridge: Cambridge University Press, 1988), book II, s. 27, p. 288, my emphasis.

46 ibid., my emphasis.

47 R. Nozick, *Anarchy, State, and Utopia* (Oxford: Basil Blackwell, 1974), pp. 175–82.

48 H. Steiner, 'Nozick on appropriation', *Mind* 87 (1978), 109–10.

49 However, there is a sense in which the societal claim, which I consider below, is what lies behind the proviso. I am grateful to Hillel Steiner for pointing this out to me.

50 Temporary secrecy may, in fact, be desirable. Cf. Bok Sissela, *Secrets – On the Ethics of Concealment* (Oxford: Oxford University Press, 1986), p. 160:

> To the extent that scientists fear theft and anticipation of their ideas, secrecy counteracts these fears and allows leeway for undisturbed research.

51 One commercial system which has much to recommend it in this context is described in C. A. Erin and J. Harris, 'A monopsonistic market – or how to buy and sell human organs, tissues and cells ethically', in I. Robinson (ed.), *The Social Consequences of Life and Death under High Technology Medicine* (Manchester: Manchester University Press, forthcoming).

52 It seems often to be forgotten that donation implies prior ownership, and hence the property status, of what is donated. Consider a legal definition of 'donation' (*Chambers English Dictionary* (Cambridge: Chambers, 1988)):

> The act by which a person freely transfers his title to anything to another.

53 From Barinaga, op. cit. at n. 12:

> Golde agrees consent forms to cover research on removed organs will probably become standard. Furthermore, he warns, the consent procedure may give many patients the false impression that they are walking gold mines. Golde says he has already received a letter from

a woman with hairy cell leukaemia who was shopping around for a researcher who is willing to buy her spleen.

54 J. Harris, 'In vitro fertilization: the ethical issues', *Philosophical Quarterly* 33 (1983), 231.

55 Such 'compulsory purchase' is discussed further in Erin and Harris, op. cit.

56 For an interesting discussion of the distinction between (moral) wronging and (legal) harming, see J. Harris, 'The wrong of wrongful life', *J. Law & Society* 17 (1990), 90–105.

10

WHAT 'BUGS' GENETIC ENGINEERS ABOUT BIOETHICS

The consequences of genetic engineering as post-modern technology

Peter R. Wheale and Ruth M. McNally

> Maybe the most certain of all philosophical problems is the problem of the present time, and of what we are, in this very moment.
>
> (Foucault 1982: 216)

INTRODUCTION

The term 'modernity' was introduced by scholars to describe the emergence of the new age of industrial society which began around the mid-seventeenth century. It attempts to capture the notion of newness and of historical discontinuity (see, for example, Frisby 1985). As a concept, modernity has the dual purpose of both attempting to describe an historical event and acting as a heuristic device (Abrams 1982) to help the student of society to gain greater understanding of industrialism.

In this paper we use Anthony Giddens's characterization of late modernity as presented in his book *The Consequences of Modernity* (1990) as a heuristic device with which to examine recombinant DNA technology – the techniques which underpin modern genetic engineering. We argue that recombinant DNA technology has an important function in the transformation process of the institutions of modernity into those of a post-modern order. We conclude that through a process of continuous reflexivity brought about by both the communicated promises of its supporters and the criticisms of its opponents it is playing a leading role in transforming the institutional dimensions of late modernity along the contours of a utopian post-modern order.

In section II we give a brief historical account of genetic engineering in order to illustrate the essentially 'discontinuous' nature of recombinant DNA technology. In section III we describe the four institutional dimensions of modernity as perceived by Giddens (1990) and the way in which they can be applied to analyse modern genetic engineering. Section IV deals with the sources of dynamism which lie behind these institutional dimensions and the way in which they characterize modern genetic engineering. Section V explores trust and risk as characteristics of modernity and their importance for the social relations of genetic engineering. In section VI we argue that the reflexivity created by the communications of social movements responding to the high-consequence risks of modernity has a dynamic function in the process of institutional transformation from modernity to post-modernity. In section VII, in regard to modern genetic engineering, we conclude that both the demands of its critics and the promises of its proponents converge on a future orientated along the contours of a utopian yet realizable post-modern order.

II GENETIC ENGINEERING

Genetic engineering is the manipulation of heredity or the hereditary material, and its aim is to alter cells and organisms so that they produce more or different chemicals or perform better or new functions. There are only three known basic ways of altering heredity: by mutation – changing the hereditary material itself; by hybridization – combining the hereditary material from two different sources; and by selection – the selection of hereditary material for the next generation (see, for example, Sober 1984). Each of these three ways of altering hereditary forms part of the process of evolution by natural selection as proposed by Charles Darwin in the nineteenth century (Darwin 1859).

Genetic engineering is a pre-modern activity in the sense that people have tampered with heredity for as long as they have cultivated crops and bred livestock, and are responsible for countless alterations of the inherited properties of life forms on the planet. However, in this century, the science of genetics has transformed the traditional craft of genetic engineering into a modern science-based technology. The basic principles of classical genetics, which were developed between 1910 and 1940, changed the study of inheritance from a descriptive, anecdotal account of various hybrid crosses to a rigorous science (see Stent 1971). Application of the principles of classical genetics had a profound effect on crop plant and domestic animal breeding, as 'rule-of-thumb' breeding procedures were replaced with rational regimes of artificial selection and hybridization, and artificial *in vivo* mutagenesis was used to generate genetic variation (see, for example, Magner 1979; Wheale and McNally 1988a: chapters 1 and 2).

Scientific breakthroughs in microbiology, biochemistry and molecular biology since the Second World War created what is often now referred to as the 'molecular revolution' in the science of genetics, and it is molecular genetics which underpins recombinant DNA technology – a radically new set of techniques which were developed in the early 1970s (see, for example, Gribbin 1985; Wheale and McNally 1988a: chapter 2; Sylvester and Klotz 1983). Recombinant DNA technology enables genetic engineers to decode, compare, construct, mutate, excise, join, transfer and clone specific sequences of DNA. It is this *micro*genetic engineering, as we have called it elsewhere (see Wheale and McNally 1986; 1988a: chapter 2), which distinguishes late twentieth-century genetic engineering from earlier genetic engineering methods, and enables the manipulation of genetic material by intervention at the molecular level rather than merely at the level of whole animals and plants.

III THE INSTITUTIONAL DIMENSIONS OF MODERNITY

Giddens postulates that there are four key interacting institutions which characterize modernity. The conceptions of the first three – capitalism, industrialism and surveillance – are substantially informed by the works of the founding fathers of sociology, namely Marx, Durkheim and Weber, and the fourth, which they comparatively neglected, as Giddens observes, is military power.

Capitalism he defines as a system of commodity production, centred upon relations between private ownership of capital and propertyless wage labour (1990: 55); industrialism as the use of economic resources and mechanization to produce goods (1990:

Figure 10.1 The institutional dimensions of modernity

56); surveillance as the supervision of the activities of subject populations in the political sphere and elsewhere; it may be direct or, more characteristically, indirect and based upon the control of information (1990: 58); and military power as the control of the means of violence by the nation-state and the 'industrialization of war', as a central institutional feature of modernity (1990: 58).

Figure 10.1 sets out these four basic institutional dimensions of modernity and their interrelations according to Giddens (1990). Capitalism, he suggests, involves the insulation of the economic markets (the 'commodification' of labour, for example) from politicalization. Surveillance is closely associated with the control exercised by modern nation-states and with military operations. Figure 10.1 also indicates that there are links between military power and industrialism, and between industrialism, capitalism and surveillance ('Taylorism' in the workplace, for example (see Wheale 1986)) in conditions of modernity (see Giddens 1990: 59–62).

Starting with capitalism and following Figure 10.1 in a clockwise direction, we now briefly consider these four institutional dimensions in relation to recombinant DNA technology.

1 Capitalism

The proponents of recombinant DNA technology, for example the European Commission and the interest groups representing the bio-industries (for example the SAGB in Europe and the BIA in the UK), claim that this new technology has the potential to satisfy world markets at competitive prices. They argue that major savings will accrue to consumers through improved efficiency and reduced prices of food and other consumer products, and from entirely new products and services; that the environmental applications of modern biotechnology will produce enormous savings in the economic costs of pollution; and that biotechnology offers real economic hope to Europe's poorer regions (see, for example, SAGB, 1990: 5–11; SEC (91) 629: 3).

Recombinant DNA technology has facilitated the entry of transnational corporations from the agri-petro-chemicals industry (including Bayer, Ciba-Geigy, Rhone-Poulenc, Monsanto, Hoechst and Du Pont) into plant breeding, for example, thus creating the conditions conducive to economic concentration of this business. In an attempt to insulate the bio-industries from the political these corporations and their industrial interest groups are actively seeking the support of the regulatory authorities to deregulate the use of recombinant DNA technology (see Triano and Watzman 1991) as they consider government controls are overburdening industry with constraints and deterring investment in biotechnology.

2 Surveillance

Recombinant DNA technology has been used to develop a range of devices for the most subtle surveillance of individuals and groups. In the medical field genetic screening facilitates the definition of what Foucault (1982) refers to as 'normalized' subjects, for example the concept of a genetically healthy baby. Genetic sceening tests and 'DNA fingerprinting' are now being developed, not only for detecting single-gene disorders and carriers of the deleterious genes deemed responsible, but for multi-factorial conditions such as schizophrenia, criminality and anti-social behaviour. The police authorities, insurance firms, financial institutions and employers are anxious to avail themselves of the new DNA fingerprinting, screening and diagnostic technologies.

Our social relationships are likely to be transformed by the knowledge flowing from human genome research, which aims to sequence all functioning DNA typically present in the human organism. According to Reilly (1977), for example, a socially conscientious health system would include a national registry of genetic profiles derived from genetic screening tests performed routinely at birth and fed into a computer data bank. Preliminary scanning of genetic profiles would identify individuals requiring immediate therapy. Under such a scheme, couples applying for marriage licences would have their respective genetic profiles analysed to reveal which genetic traits they might be both carriers of in order to identify which genetic disorders their children might be at risk of being afflicted with. Parental options already include genetically screened eggs and sperm as alternative sources of genetic material if their own is considered defective (see Wheale and McNally 1988a: chapter 10).

3 Military power

The agents of biological warfare are bacteria, viruses and fungi, and the target organisms may be people, livestock or crops. Recombinant DNA technology has revolutionized our ability to construct microbes with previously unattainable combinations of characteristics, which has added another dimension to arms production. The new potential of biological warfare has revitalized both defensive and offensive biological bio-weapons research despite international conventions and treaties which ban the development of biological weapons (see Wheale and McNally 1988a: chapter 8).

4 Industrialism

The two major components of a successful industrial policy are the development of technological innovations of great utility and the presence of sufficiently large markets for these innovations and the goods and services derived from them to sustain production cycles sufficient to provide a commercial return on the capital investment required for their implementation. The newly emerged bio-industries manufacture and supply the enabling technological innovations for genetic engineering research and biotechnology (see, for example, Bull *et al.* 1982; Fishlock 1982).

These technological innovations include automated DNA synthesizers and decoders, computerized data banks of DNA/RNA and protein sequences, biosensors, immobilization systems, and automation and computerization systems for fermentation (see Wheale and McNally 1986). Recombinant DNA technology converging with a cluster of innovations including computerized automation, electronics and laser technology has formed a pattern of change so economically important as to constitute a 'new technological system' (Clark *et al.* 1984).

Underpinned by recombinant DNA technology, this new technological system is already having profound affects upon our social and physical environment. It is a world-wide complex of scientific expertise, technological capabilities and transnational capital accumulation operating in international markets. It interfaces with all sectors of society – industrial sectors include chemicals, food, agriculture, energy, resource-recovery, environmental control, health care and pharmaceuticals; and these in turn interface with the military and with numerous governmental agencies – on a global scale, comprising what the authors have called elsewhere the 'bio-industrial complex' (see Wheale and McNally 1988a: part II).

IV THE THREE SOURCES OF THE DYNAMISM AND THE GLOBALIZING SCOPE OF MODERN INSTITUTIONS

In Giddens's view behind these institutional clusterings of capitalism, industrialism, surveillance and military power lie the three sources of dynamism of modernity. These are as follows: time–space distanciation; disembedding mechanisms; and reflexivity (Giddens 1990: 63). In this section we shall illustrate the operation of these suggested sources for the dynamism of modernity using genetic engineering as a case-study.

1 Time–space distanciation

Time–space distanciation is 'the separation of time from space and their recombination in forms which permit the precise time–space "zoning" of social life' (Giddens 1990: 16–17).

This particular source of the dynamism of modernity derives from technological innovation. Scientific breakthroughs in microbiology, biochemistry and molecular biology since the Second World War formed the basis of the molecular revolution in genetics. This revolution underpins a new set of techniques, significantly, in this context, called *recombinant* DNA technology. And it is recombinant DNA technology which enables modern genetic engineers to recombine genetic material, thereby altering the time–space relationships of nucleic acids, genomic ecosystems, mobile genetic elements, and genetically engineered organisms.

Spatial dislocation is brought about by the manipulation of nucleic acids outside of living cells and their reinsertion into foreign cells.

Genetically recombinant organisms are also *temporally* dislocated – that is, dislocated from their own genesis (through somatic cell manipulation) and from their own heredity (through germ-line manipulation). The ontological framework of biology rests upon measurements of the passage of time. Time is measured by change, and change – differentiation and development, senescence and death, adaptation and speciation – is an essential characteristic of life. For example, the concept of species: according to conventional evolutionary theory, a species is a unit of biological classification which is reproductively separated from other units of biological classification through functions of space and time.

Recombinant DNA technology breaches the temporal divide of genetic incompatibility; it reduces the time needed to produce new varieties; and it alters the temporal relations of genesis: a unique recombinant organism can be cloned – that is, spatially divided into many individuals – and with cryogenesis individuals from the same clone may be separated from each other temporally and 'born' into different generations.

2 Disembedding mechanisms

Disembedding mechanisms are 'mechanisms which prise social relations free from the hold of specific locales, recombining them across wide time–space distances' (Giddens 1991: 2). These disembedding mechanisms are abstract systems, and Giddens (1990; 1991) distinguishes two types of disembedding mechanisms – symbolic tokens and expert systems – which are intrinsically involved in the development of modern social institutions. Genetic engineering is such a disembedding mechanism – both as a symbolic token and as an expert system:

(a) Nucleic acid sequence data as a symbolic token

Symbolic tokens are media of interchange which can be 'passed around' without regard to the specific characteristics of individuals or groups which handle them at any particular juncture, for example money (Giddens 1990: 22).

The entire science of genetics is based upon an information metaphor. Since the elucidation of the genetic code, it is common for nucleic acid molecules to be characterised and denoted in terms of the sequence of their bases. Thus rendered as pure information, the three-dimensional molecules of DNA and RNA are reduced to two-dimensional linear symbols, and transformed into symbolic tokens, a semiotics which constitutes the international *lingua franca* of molecular biology.

(b) The bio-industrial complex as an expert system

Expert systems are 'systems of technical accomplishment or professional expertise that organise large areas of the material and social environments in which we live today' (Giddens 1990: 27).

The 'bio-industrial complex' (see above) has generated a global expert system whose technical accomplishments affect our social and physical environment. Expertise generated by this system includes environmental risk-assessment, public health and consumer safety, genetic diagnostics, counselling services and therapy for a wide range of conditions. In a similar way to other expert systems, its products and services are used by consumers who lack full information about the products and services they use and who therefore must place trust in the expertise of the system (see below).

3 Reflexivity

Reflexivity is the reflexive ordering and reordering of social relations in the light of continual inputs of knowledge affecting the actions of individuals and groups (Giddens 1990: 17).

Recombinant DNA technology can be used to construct knowledge about human populations and individuals. Genetic knowledge of populations constitutes a new system of classification. It is knowledge which fixes the 'normal' – it provides a new standard of normalcy with which individuals may both be judged and judge themselves and thereby changes their objectivities and their subjectivities. For example, the genetic screening expert system is constructing genetic disorders as a problem:

(a) for the families and individuals affected by such disorders;
(b) for society – the economic burden of genetic disease.

The discourse on the problem of genetic disorders and its constructed solution – pre-natal diagnosis and selective termination – will alter both the subjective states and objective statuses of individuals and the expert systems themselves. It will change the subjectivities of prospective parents – 'responsible parents' will be socially constructed as parents who choose to have pre-natal diagnosis and choose to have a therapeutic abortion of affected foetuses. It will change attitudes towards those with genetic diseases, since such people have been constructed as the problem – the burden of genetic disorders. It will generate expectations that the genetic screening expert system can solve 'the problem'.

Recombinant DNA technology also challenges our beliefs about evolution and the relative status of our species in the evolutionary process. The expression of human genes in bacteria, pigs and tobacco plants challenges philosophies which grant human beings a privileged position among species, and which defend 'speciesism' on the basis that humans are fundamentally different from other species. Ryder (1990), for example, asks provocatively: 'How many human genes would warrant that a recombinant genetically engineered pig be treated as human being?'

V TRUST AND RISK

Time–space distanciation, disembedding mechanisms and reflexivity are each a source for the dynamic transformation of modern institutions, and each of these three sources is inherently globalizing. Moreover, these three sources of the dynamism and globalization of modern institutions give rise to two other characteristics of modernity, namely trust and risk (Giddens 1990: 63), and it is these two characteristics which we now consider.

1 Trust

To have complete confidence in technical accomplishment or professional expertise one would be required to make one's own judgment based on complete information. In Giddens's view the counterfactual, future-orientated character of modernity is largely structured by trust vested in abstract systems (1990: 84): 'all disembedding mechanisms, both symbolic tokens and expert systems, depend upon trust' (1990: 26) because they 'remove social relations from the immediacies of context' (1990: 28). If we accept this viewpoint, then we must accept that trust is involved in a fundamental way with the institutions of modernity. As Giddens (1990) asserts, if faith in the epistemology used by the expert system declines, then the public's trust in it cannot be sustained, and the outcome is public anxiety.

Communication, according to Luhmann (1986; 1988) generated from any given *a priori* assumption stabilizes our normative expectations because it is self-referential, and the truth claims are symmetrical with the *a priori* assumption. A communication system is also autonomous in the sense that it is normatively closed: it does not make reference to external *a priori* assumptions. According to Luhmann (1986) communication systems change and develop through the input of cognitive elements which involve three processes: variation from individuals as 'psychic systems', the process of selecting cognitive inputs and stabilization of the autopeisis of the social system. It is through the repetition of specific communications that normative expectations are established in society. These normative expectations reduce perceived uncertainty and, where they are generalized expectations, function as social structures. However, when alternative communications about the same informational set are made, then expectations will be fragmented.

Despite their apparent success, modern genetic engineers are often surprised by the lack of public trust in their endeavours, and, in general, are dismayed by the extent to which they are called upon to justify their experiments and innovations (see Denselow 1982; Gershon 1983; Krimsky 1982). We believe the reason for this public scepticism is that as the public becomes increasingly cognizant of the ungrounded nature of scientific epistemology and the contingency of scientific knowledge (Barnes and Edge 1982; Mulkay 1979; Lakatos and Musgrave 1970), the status of scientific communications is being reduced. The idea that knowledge is contingent and revisable, that there is no 'totalising self-referential critique', no way of ultimately 'distinguishing between theory and ideology' or between 'true consensus' and 'false consensus' (Habermas 1987), is undermining the legitimacy of traditional scientific authority, and this is reflected in the public's attitude to scientific expert systems. When there is a loss of faith in the epistemology which underpins *all* science-based expert systems, then the public can no longer trust any expert system to appraise it of the dangers and risks of technology because the theory underpinning the identification and measurement of these is underdetermined, that is, ultimately unknowable.

2 Risk

As trust in technology and scientific expertise is closely associated with the notion of 'acceptable risk', and as the latter cannot be firmly established (because, as we have observed above, no scientific epistemology exists which could provide tests which accurately measure it), the social relations of science in general (see, for example, Gibbons and Wittrock 1985), and of genetic engineering in particular, are rapidly deteriorating. Public scepticism regarding the claims made by the expert systems such as that of the 'bio-industrial complex' concerning the safety, utility and public desirability of modern science and technology is believed to be destabilizing the relationship between the expert and the lay public.

One important consequence of this is that sectional interests which rely on the traditional authority of science can now be challenged with greater conviction by alternative knowledge claims (for example those of 'alternative medicine', 'alternative food' and 'green consumerism'), thus creating a dialectic (reflexive rather than materialist) in advanced modern culture.

The individual's perception of risk in modern society is an important factor for social action. Modernity is certainly a risk culture (Giddens 1991: 3). But this is not to say that 'social life is inherently more risky that it used to be; for most people in the developed societies, that is not the case' (Giddens 1991: 3). But the *intensity* of risk is surely the basic element in the 'menacing appearance' of the circumstances in which we live today (Giddens 1990: 125). The high-consequence risks of modernity as suggested by Giddens (1990: 171) are depicted in Figure 10.2 as essentially superimposed on the four institutional dimensions of modernity.

As a source of dynamism and globalization – in particular through time–space distanciation – modern genetic engineering has the potential to be a major contributor to the high-consequence risks of modernity. Below we describe some examples of the high-consequence risks of genetic engineering.

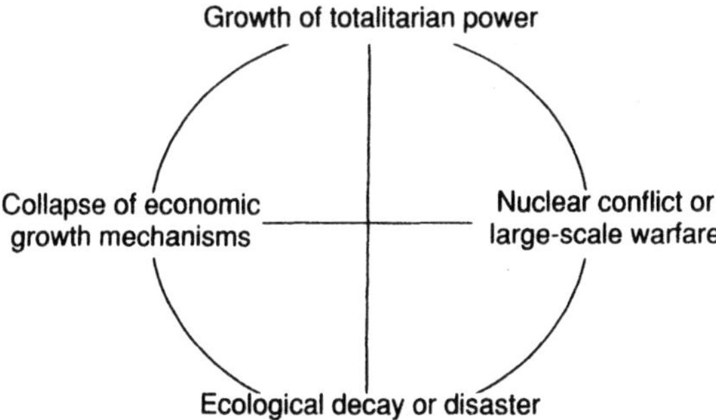

Figure 10.2 High-consequence risks of modernity

(a) Capitalism – patentability of recombinant products and processes

The high-consequence risk through capitalism is the collapse of global economic exchange. Recombinant DNA technology is uniquely able to contribute to this process of economic collapse through the patenting of life forms.

Until the advent of recombinant DNA technology, patenting legislation all over the world explicitly excluded the patentability of plant and animal varieties and the biological methods and processes used for their production on the grounds that they were not inventable. With the advent of recombinant DNA technology, it was no longer in the interests of the bio-industries to consider the world's stock of genetic resources to be common heritage, and it has achieved a major re-evaluation of the status of evolution versus invention in the context of patenting law (see Wheale and McNally 1990a: part I).

The patentability of recombinant products and processes has proved to be a powerful economic incentive to substantial corporate investment in molecular biology, for example in human genome research. Radical technological change gains some of its impetus from the imperatives of capitalist accumulation and civil and military procurement, but once underway, many believe it displays a dynamism of its own (see, for example, Ellul 1965; Marcuse 1972; Giddens 1990). However, in the case of recombinant DNA technology its technical merit is now *less* important to the bio-industrial complex than the economic power which it confers through the patentability of the world's genetic resources.

(b) Industrialism – biogenetic waste, loss of biodiversity and pandemics

Examples of the high-consequence risks of industrialism are ecological calamity and uncontrollable population explosion.

Through recombinant DNA technology, more organisms of novel genetic composition are now being produced and introduced – accidentally and deliberately – into the environment than would ever have been possible in an equivalent period of time using conventional breeding practices. Recombinant organisms can reproduce, migrate and mutate and could become a growing, moving, changing form of pollution, which we have called elsewhere 'biogenetic pollution' (see Wheale and McNally 1988a: chapter 7). The high-consequence environmental risks of recombinant organisms are a result of their potential to: compete successfully with existing species; have beneficial effects on undesirable species; and change physical factors, such as weather patterns.

Not only is recombinant DNA technology the source of environmental risk, but

> it can be harmful to animal welfare, both by design – human disease models – or by accident, or because of the toll of increased farming efficiency;
>
> it has the innate tendency for genomic disruption – an appreciable increase in the number of novel mobile genetic elements, which will occur as they are exploited as vectors in recombinant DNA technology, increases the probability of initiating major genomic alterations with consequences for cells, organisms, populations and species (see Wheale and McNally 1986; 1988a: chapter 4);
>
> and it has the potential to create new epidemics, particularly viral epidemics. Viral vectors could themselves mutate to virulent form; they may recombine with other viruses and genetic elements resulting in virulent new strains; or their integration into the host genome could trigger the liberation of virulent viruses that were previously dormant (see Wheale and McNally 1986; 1988a: chapter 4; NAVS 1987).

(c) Surveillance – Bio-informatics

The high-consequence risk of surveillance is 'the intensifying of surveillance' for political power for totalitarian purposes (Giddens 1990: 172).

Nucleic acid sequences provide information about our past, present and future, both as individuals and as a species. Nucleic acid sequence data lend themselves to applications which require the classification of people into groups for the purpose of institutional decision-making. Biometrical genetics research is already being carried out to establish correlations between genetic constitution and intelligence, anti-social behaviour, mental illness and workplace disorders (see Rose *et al.* 1984). Indeed, bio-informatics – computerized nucleic acid sequence information – is a burgeoning industry in its own right. There already exist data banks of personal genetic profiles gathered for health monitoring, reproductive choices, forensic purposes, immigration and paternity disputes, life assurance contracts, and pre-employment genetic screening. These data, which can be transmitted electronically in the same way as electronic mail, have enormous implications for surveillance, both public and private, which is considered to be a form of administrative control (see Foucault 1977; Turkel 1990).

(d) Military power – biological weapons

The post-Second World War trend towards biological warfare disarmament and the banning of biological weapons – most notably expressed in the 1972 Biological Weapons Convention – was short-lived as a direct result of the advent of recombinant DNA technology.

Theoretically, recombinant DNA technology can be used to make biological weapons which are more reliable and predictable than previously. Recombinant DNA technology reinstated biological weapons as a potential threat to national security, and it is this threat which has been used to justify an increase in military research to develop new recombinant vaccines and anti-toxins against biological weapons.

It is because the distinction between offensive research and defensive research is not clear-cut that many scientists and public-interest groups are concerned that defensive biological warfare research is being used as a 'Trojan horse' to undertake military research using genetic engineering which is actually for offensive purposes (see Wheale and McNally 1988a: chapter 8), for example to design recombinant vaccines with which to protect a state's own troops from the biological weapons it intends to use itself on an enemy power. The high risk is that the effects of a biological war

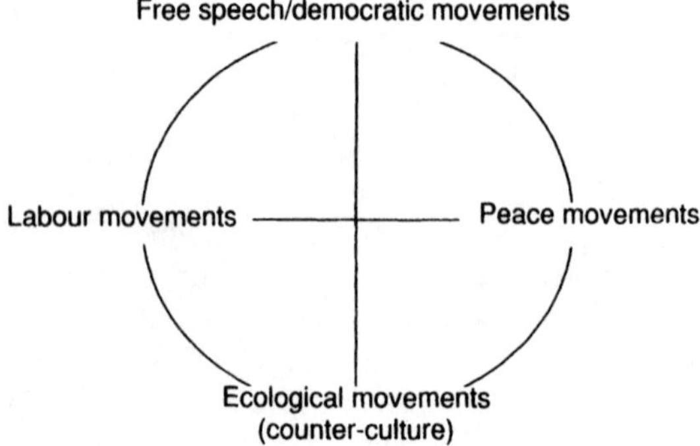

Figure 10.3 Types of social movements

could be more terrible than even those of a nuclear war, because biological weapons replicate uncontrollably and go on killing indefinitely.

VI SOCIAL MOVEMENTS

Another feature of modernity is the existence of social movements whose formation is a response to an awareness of the high-consequence risks which modern institutions engender. Figure 10.3 indicates the types of social movements characteristic of modernity which are essentially superimposed on the four institutional dimensions of modernity (Giddens 1990: 159).

The realization that empirical knowledge does not provide us with a decision-rule to decide between different value positions, together with acceptance that scientific knowledge lacks foundationalism, is providing a social environment receptive to alternative knowledge claims from social movements such as conservationists, environmentalists, animal welfarists and feminists.

Counter-culture expert systems are generating rival knowledge claims about values and risks based on different *a priori* assumptions from those of scientific expert systems. However, the knowledge claims of bioethics can be reflexively appropriated by scientific expert systems. For example, the bio-industrial complex has incorporated some of the ecological discourse concerning environmental risk into its communications (see, for example, Bishop 1990). It is suggested that the reflexive appropriation and reappropriation of knowledge between expert systems and the conflict of knowledge claims is a salient feature of modernity (Lyotard 1985; Giddens 1990).

The proliferation of bioethics communications issuing from public pressure groups is an example of resistance to (bio-) power over life (see Foucault 1979) and an indication of the extent to which the public has lost trust in the rationality of scientism and is disenchanted with the consequences, intended or unintended, of many of the activities of the 'bio-industrial complex'.

Modern genetic engineering occupies a pivotal position in the institutions of modernity, and it engenders high-consequence risks. It continues to raise questions concerning our rights and responsibilities towards other species, future generations and the very survival of the planet (see, for example, Holland 1990). DNA profiles, for example, impinge upon reproduction, employment, immigration and forensic science. Recombinant DNA technology produces potential biohazards in the workplace and for consumers of food and health care, and it has frightening military applications.

Recombinant DNA technology is implicated in virtually every major ethical controversy preoccupying modern society, including issues of: animal welfare, the environment, natural rights and speciesism, religion, evolutionary theory, social control, eugenics, the North–South divide, control over natural resources, biological warfare, food and farming, and medicine and agriculture disputes.

Since the advent of *micro*genetic engineering, starting with the recombinant DNA debate in the 1970s, to the present – when the number of public-interest groups devoted to genetic engineering issues or with genetic engineering on their agenda increases daily – modern genetic engineering has stimulated social movements which articulate what they consider to be the unacceptable risks of this new technology and prescribe guidelines for alternative future transformations (see Wheale and McNally 1990b).

Giddens postulates that any post-modern system will be institutionally complex with a plurality of agencies. We can characterize it as representing a movement 'beyond' modernity along each of the four institutional dimensions of modernity (Giddens 1990: 163). Future possible transformations constitute the contours of a (realizable) utopian post-modern order (Giddens 1990: 164) as indicated by Figure 10.4, and also a dystopian future brought about because of the high-consequence risks which exist in

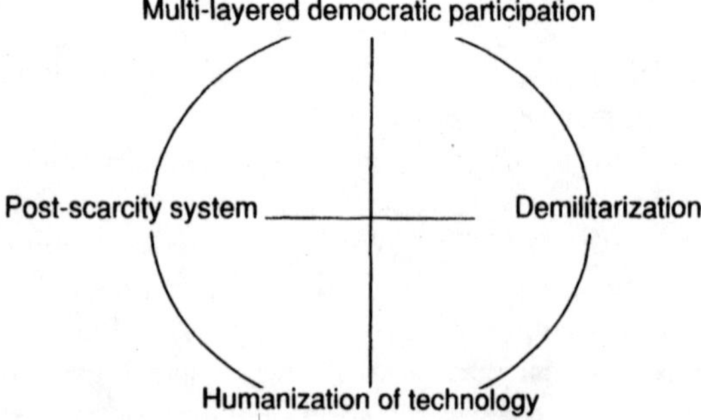

Figure 10.4 The contours of a post-modern order

radicalized modernity as discussed above.

Table 10.1 briefly indicates the guidelines for the transformation by modern genetic engineering of the afore-mentioned four modern institutions towards a utopian post-modern order generated by new 'alternative' social movements.

VII CONCLUSIONS

It remains a controversial question whether the advanced industrial societies are on the threshold of a post-modern age (see, for example, Bell 1973; Lyotard 1985: 14; Lash and Urry 1987). Giddens (1990) considers that, as long as the four basic institutional dimensions of modernity prevail – capitalism, surveillance, military power and industrialism – in their present form, we are still in modernity. However, he concedes that we are moving into an age in which the consequences of modernity are becoming more radicalized and universalized than before' (1990: 3; see also 51). He suggests that: 'this means that the trajectory of social development is taking us away from the institutions of modernity towards a new and distinct type of social order' (Giddens 1990: 46).

Giddens's (1990) schema of modernity and its consequences takes us, however, at least to the point where social movements are making guidelines for the future transformation of modern institutions, guidelines which define the contours of a post-modern order. We have described what he suggests the contours of this new and distinct type of social order might be. Using Giddens's conjectures as a heuristic device for our analysis, we have argued that recombinant DNA technology is playing an important role in

Table 10.1 Genetic engineering's role in alternative contours of post-modernity

1 CAPITALISM
Utopian: A post-scarcity system in which greater value is placed on *stewardship* rather than ownership and exploitation of genetic resources. Dystopian: Collapse of economic growth mechanisms by the appropriation of the world's genetic resources by the bio-industrial complex using patenting.
2 INDUSTRIALISM
Utopian: A system of planetary care through sustainable development. A future in which the human species upholds its responsibility towards the welfare and survival of other species and the long-term future of the planet. Dystopian: Ecological decay or disaster caused through biogenetic waste, loss of biodiversity and pandemics.
3 SURVEILLANCE
Utopian: Multi-layered democratic participation in decisions over the collection, storage and use of genetic data. Dystopian: A growth in totalitarian power, a rise in genetic determinism and eugenics movements and the use of personal genetic profiles to discriminate against individuals and groups, for example, in employment, insurance, immigration and reproduction.
4 MILITARY POWER
Utopian: The transcendence of war through demilitarization achieved by international peace treaties, the banning of biological research by the military, and the institutionalization of 'responsible science'. Dystopian: Uncontrolled proliferation of biological weapons, for example viruses, bacteria or fungi, resulting in disease and death of livestock, crops and/or people.

transforming the four institutions of modernity into a post-modern order. We consider that the dynamic factor in this process is the social movements responding to recombinant DNA technology. It is the guidelines which they communicate which are themselves an important source of dynamism which alter not only the reflexive process itself but the other sources of dynamism, namely time–space distanciation and the disembedding mechanisms discussed above.

This last observation gives us grounds for reasoning that the normative expectations of expert systems change over time, especially when they reflexively appropriate alternative (i.e. counter-culture) knowledge claims. In the case of the 'bio-industrial complex', we have argued that both the critics' demands and the proponents' promises converge on a future orientated along the contours of a realizable utopian post-modern order. For example, consider the following trends which result from pressures from public-interest groups concerned about the high-consequence risks of recombinant DNA technology:

> the potential for time–space distanciation has been curtailed by regulations which prohibit certain forms of genetic recombination which are considered to present 'unacceptable' risks;
>
> new expert systems – expert systems which are trusted to evaluate risk – are now being formed, for example: bioethics committees; recombinant DNA advisory committees; animal experimentation and farm-animal welfare committees; and environmental risk-assessment committees;
>
> the 'greening of biotechnology' is occurring; recombinant DNA technology has been redefined as the *solution* to the high-consequence risks of modernity – both to the risks that it engenders and to the risks arising from other technologies and disembedding mechanisms.

The above examples illustrate the reflexivity operating on the 'bio-industrial complex'. In our opinion, this reflexive process is tending to adapt existing institutional norms to more humanistic and ecologically based values which are responsibly constrained by a future-orientated ethics (see, for example, Brundtland 1987; Jonas 1984). The social relations of the 'bio-industrial complex' will surely improve once the general public perceives that it is acting in such a way as to preclude a future of unsustainable resources depletion, global pollution, industrialized militarism and totalitarian control.

REFERENCES

Abrams, P. (1982) *Historical Sociology*, London: Open Books.

Barnes, B. and Edge, D. (1982) *Science in Context*, Milton Keynes: Open University Press.

Bell, D. (1973) *The Coming of Post-Industrial Society: A Venture in Forecasting*, New York: Basic Books.

Bishop, D. (1990) 'Genetically engineered insecticides: the development of environmentally acceptable alternatives to chemical insecticides', in P. R. Wheale and R. McNally (eds) *The Bio-Revolution: Cornucopia or Pandora's Box?*, London: Pluto, pp. 115–34.

Brundtland G. H. (1987) *Our Common Future*, Paris: UN Commission on Environment and Development.

Bull, A. T. *et al.* (1982) *Biotechnology: International Trends and Perspectives*, Paris: OECD.

Clark, J. *et al.* (1984) 'Long waves, inventions, and innovations', in C. Freeman (ed.) *Long Waves in the World Economy*, London: Frances Pinter.

Darwin, C. (1859) *On the Origin of Species by Means of Natural Selection or the Preservation of Favoured Races in the Struggle for Life*, London: John Murray; reprinted Cambridge, Mass.: Harvard University Press, 1975.

Denselow, J. (1982) 'GMAG and the teenage jackass', *New Scientist*, 26 August: 558–61.

Ellul, J. (1965) *The Technological Society*, London: Cape.

Fishlock, D. (1982) *The Business of Biotechnology*, London: Financial Times Business Information.

Foucault, M. (1977) *The Archaeology of Knowledge*, London: Tavistock, p. 138.

—— (1979) *The History of Sexuality*, vol. 1: *An Introduction*, London: Allen Lane, Penguin.

—— (1982) 'The subject of power', in H. L. Dreyfus and P. Rabinow, *Michel Foucault: Beyond Structuralism and Hermeneutics*, Hemel Hempstead: Harvester Press.

Frisby, D. (1985) *Fragments of Modernity*, Cambridge: Polity Press.

Gershon, S. (1983) 'Should science be stopped? The case of recombinant DNA research', NIMH, *Public Interest* 71: 3–16.

Gibbons, M. and Wittrock, B. (eds) (1985) *Science as a Commodity*, Harlow: Longman.

Giddens, A. (1990) *The Consequences of Modernity*, Cambridge: Polity Press.

—— (1991) *Modernity and Self-Identity*, Cambridge: Polity Press.

Gregory, G. (1984) 'Big is beutiful in biotechnology', *New Scientist* 104 (1433): 12–15.

Gribbin, J. (1985) *In Search of the Double Helix*, London: Corgi.

Habermas J. (1984) *The Theory of Communicative Action I: Reason and Rationalization of Society*, Boston: Beacon.

—— (1987) 'Questions and counterquestions', in R. J. Bernstein (ed.) *Habermas and Modernity*, Cambridge: Polity Press, pp. 192–216.

Holland, A. (1990) 'The biotic community: a philosophical critique of genetic engineering', in P. R. Wheale and R. McNally (eds) *The Bio-Revolution: Cornucopia or Pandora's Box?*, London: Pluto, pp. 166–74.

Jonas, H. (1984) *The Imperative of Responsibility: In Search of Ethics for the Theological Age*, Chicago and London: University of Chicago Press.

Krimsky, S. (1982) *Genetic Alchemy: The Social History of the Recombinant DNA Controversy*, Cambridge, Mass.: MIT Press.

Lakatos, I. and Musgrave, A. (eds) (1970) *Criticism and the Growth of Knowledge*, Cambridge: Cambridge University Press.

Lash, S. and Urry, J. (1987) *The End of Organised Capitalism*, Cambridge: Polity Press.

Luhmann, N. (1986) 'The autopoiesis of social systems', in F. Geyer and J. van der Zouen (eds) *Sociocybernetic Paradoxes: Observation, Control and Evolution of Self-Steering Systems*, London: Sage.

—— (1988) 'The unity of the legal system', in G. Teubner (ed.) *Autopoietic Law: A New Approach to Law and Society*, Berlin/New York: De Gruyter.

Lyotard, J. F. (1985) *The Post-Modern Condition: A Report on Knowledge*, Minneapolis: University of Minnesota Press.

McNally, R. and Wheale, P. R. (1986) 'Recombinant DNA technology: re-assessing the risks', *STSA Newsletter*, summer: 56–69.

—— (1988) *Deliberate Release into the Environment of Genetically Engineered Organisms*, Brussels: GRAEL, European Parliament.

Magner, L. N. (1979) *A History of the Life Sciences*, New York and Basle: Marcel Dekker

Marcuse, H. (1972) 'Industrialization and capitalism in the Work of Max Weber, in H. Marcuse, *Negations*, London: Penguin.

—— *One-Dimensional Man: The Ideology of Industrial Society*, London: Routledge & Kegan Paul.

Mulkay, M. J. (1979) *Science and the Sociology of Knowledge*, London: Allen & Unwin.

NAVS (1987) *Bio-Hazard*, London: NAVS.

OECD (1986) *Recombinant DNA Safety Considerations: Safety Considerations for Industrial, Agricultural and Environmental Applications of Organisms Derived by Recombinant DNA Techniques*, Paris: OECD.

RCEP (1989) *The Release of Genetically Engineered Organisms to the Environment*, London: HMSO.

Reilly, P. (1977) *Genetics, Law and Social Policy*, Cambridge, Mass.: Harvard University Press.

Rose, S. *et al.* (1984) *Not in our Genes*, Harmondsworth: Penguin.

Ryder, R. (1990) 'Pigs *will* fly', in P. R. Wheale and R. McNally (eds) *The Bio-Revolution: Cornucopia or Pandora's Box?*, London: Pluto, pp. 189–94.

SAGB (1990), *Community Policy for Biotechnology: Economic Benefits and European Competitiveness*, Brussels: CEFIC.

SEC (91) 629 [The Bangemann Report 1991] *Promoting the Competitive Environment for the Industrial Activities Based on Biotechnology within the Community*, Brussels: Commission of the European Communities.

Sober, E. (ed.) (1984) *Conceptual Issues in Evolutionary Biology*, Cambridge, Mass., and London: MIT Press.

Stent, G. S. (1971) *Molecular Genetics: An Introductory Narrative*, San Francisco: Freeman.

Sylvester, E. J. and Klotz, L. C. (1983) *The Genetic Age: Genetic Engineering and the New Industrial Revolution*, New York: Charles Scribner's Sons.

Triano, C. and Watzman, N. (1991), *All the Vice-President's Men: How the Quayle Council on Competitiveness Secretly Undermines Health, Safety, and Environmental Programs*, Washington, D.C.: OMB Watch.

Turkel, G. (1990) 'Michel Foucault: law, power, and knowledge', *Journal of Law and Society* 17 (2): 170–93.

Wheal, P. R. *et al.* (1986) *People, Science and Technology*, Hemel Hempstead: Wheatsheaf, chapter 11.

Wheale, P. R. and McNally, R. (1986) 'Patent trend analysis: the case of microgenetic engineering', *Futures*, October: 638–57.

—— (1988a) *Genetic Engineering: Catastrophe or Utopia*?, London: Wheatsheaf; New York: St Martin's Press.

—— (1988b) 'Technology assessment of a gene therapy', *Project Appraisal* 3 (4), December: 199–204.

—— (eds) (1990a) *The Bio-Revolution: Cornucopia or Pandora's Box*?, London: Pluto.

—— (1990b) 'Genetic engineering and environmental protection: a framework for regulatory evaluation', *Project Appraisal* 5 (1): 23–37.

11

CATEGORICAL OBJECTIONS TO GENETIC ENGINEERING – A CRITIQUE

Matti Häyry

When new genetic technologies are examined and assessed in the ethical framework, two general types of objection are usually presented. First, some theorists appeal to the predictable consequences of employing gene technology. The core idea of the objection is to claim that the evil probably produced by genetic engineering exceeds the benefits probably flowing from its use. This approach has been dubbed in the literature 'pragmatic' or 'consequentialist'. Second, there are theorists who reject the first approach as amoral, and argue that ethical evaluations should always proceed from purely ethical considerations. Arguments of this second type can be labelled as 'categorical' or 'deontological', and they range from complex theological accounts to simple common-sense expressions of disapproval.

Although I think that genuinely pragmatic reasons for and against gene technology are decisively important, I shall say virtually nothing about them in this paper. It is the second, deontological type of objection that interests me here, because it is in a sense more fundamental than its consequentialist rival. What I mean by this is that if deontological theorists are right, they can establish the moral status of human activities – such as genetic engineering – quite independently of the expected consequences of those activities. One valid deontological objection against gene technology would be enough to put all consequentialist moralists out of business in this field.

New genetic technologies include a variety of practices, from the manipulation of plants and animals to attempts to alter human chromosomes. The categorical and deontological arguments have, however, mostly been restricted to human genetics, and I shall focus on this aspect of the issue. There are three major forms of human genetics that have been regarded as morally dubious, and to which I shall confine my attention. These are somatic cell therapy, germ-line gene therapy, and the project to map the human genome. The medico-biological aim of these practices is to identify the genes which cause known diseases, and to cure these diseases by recombinant DNA techniques.[1] Somatic cell therapy is intended to cure only the individuals who are actually being treated, whereas germ-line cell therapy is expected to rectify hereditary disorders both in the patients themselves and in their descendants. The project to map the human genome is related to these therapeutical applications in that reliable knowledge concerning the chromosomal structure of human beings would be quite invaluable to medical personnel when diagnoses are being made.

IS GENETIC ENGINEERING DANGEROUS?

The difference between 'pragmatic' and 'categorical' arguments can be illustrated by studying one basic objection against the use of biotechnology, namely the apparently simple statement that 'genetic engineering is dangerous'. Two interpretations can be given to this objection.

First, the point of the argument may be that the genetic engineering of human beings is physically dangerous to certain identifiable individuals. For instance, many experimental animals currently lose their lives in biotechnological research. Human embryos and pre-embryos are also subjected to scientific experiments, and their lives are in similar danger. And although the embryos subjected to therapeutical gene manipulation would have the status of patients, their chances of survival would still be rather low until the treatments had developed beyond the experimental level. Even adult patients, embryo patients who survive the therapy and the offspring of these groups may be in peril, since the manipulation of genes can cause new diseases as well as cure existing ones. Finally, genetic engineering as a whole is an expensive high-tech enterprise which will possibly benefit only multinational corporations and a handful of affluent westerners who suffer from fancy ailments. If scarce medical resources are primarily allocated to the development of gene therapy, more important projects such as sanitation and social security may have to be abandoned, and vast masses of people, especially in the Third World, will be endangered.[2]

The distinctive feature of all objections based on these and similar claims is that their structure is conditional. Genetic engineering, according to the objections, is dangerous only if at least one of the claims concerning its undesirable entailments is valid. Put the other way around, this implies that unless at least one of the remarks and predictions about present and future evils is true or reasonably probable, there is no tenable objection against gene technology. It is this possibility that makes objections based on physical danger pragmatic or conditional as opposed to absolute or categorical.

Second, however, the point of the argument may also be that genetic engineering is dangerous in some moral or symbolic rather than physical sense. Many theorists and a number of lay persons seem to think that gene technology is somehow inherently and irrevocably 'immoral', either because it violates the rules set by the human community, or because it is against the higher laws of God or nature. The objections based on these ideas are genuinely categorical, since the immorality of the practice under evaluation is supposed to be intrinsic (or conceptual) and therefore beyond empirical testing.

If immorality is taken to mean deviation from rules set by humans, the claim that genetic engineering is immoral does not have much bite. Rules set by humans can be altered by humans, and therefore laws and regulations prohibiting genetic experiments and therapies do not by themselves prove anything about the ultimate ethical wrongness of these activities. On the other hand, if the objection is based on spontaneous human sentiments, the difficulty is to bridge the gap between those sentiments and critical morality. Feelings certainly influence the opinions that people have on ethical issues, but it has never been conclusively shown that feelings should be uncritically allowed to enter reflective moral judgments. The fear and suspicion people may feel towards gene technology do not, therefore, count as a valid objection against it.

If, however, immorality is taken to mean transgressions of the divine or natural laws, the matter is at once far more complicated. Those who oppose genetic engineering may say, for instance, that to interfere with the human germ-line would be 'unnatural' or 'against God's will' or an instance of 'playing God'. But what exactly do these expressions mean? Can they be translated into plain language which could be understood without prior commitment to theological or metaphysical systems?

SIX WAYS OF PLAYING GOD

In a recent article Ruth Chadwick has neatly analysed and assessed the 'playing God' argument.[3] According to her analysis, the objection that an action is wrong because it is an instance of playing God has two different meanings in two different kinds of setting. In the context of sensitive medical decision-making the point of the objection is that human beings are in no position to decide legitimately about each other's fates on the basis of quality-of-life judgments. In the context of new medical technologies, again, the crux of the argument is that actions describable as playing God can lead to disastrous and unpredictable consequences. These two aspects are both present in certain forms of genetic engineering, such as germ-line gene therapy, and it is therefore useful to take a closer look at Chadwick's account.

With regard to the decision aspect of the playing-God objection, Chadwick distinguishes three major lines of argument, two of which she finds untenable.[4] First, the wrongness of playing God can be based on the idea that it is God's prerogative to give life and to take it away. Active euthanasia, for instance, has been attacked by referring to this notion. But the problem with it is that no reasonable morality condemns doctors and nurses who do their best to save and prolong lives, although this work can, according to the above interpretation, be described as playing God. Second, the point of the objection may be that in certain matters the natural course of events should be preferred to human interference. An example of such matters is the reallocation of health through medical decisions. To kill one patient in order to save two others would be the best thing to do in crude utilitarian terms, but it would also be a hideous instance of playing God.[5] In situations like this, so the argument goes, doctors can act morally only by letting nature take its course. The obvious difficulty in this second interpretation is that whatever decisions doctors make, they cannot help playing God in the defined sense. Refusals to alter the 'natural' course of events affect the patients and their lives as much as any positive action.

Third, the formulation that Chadwick finds plausible and morally relevant is founded on the equality and limited knowledge of human beings. In matters concerning life and death we may justifiably feel that no one else is qualified to judge whether our lives are worth living. This conviction stems from two factors. On the one hand, it can be argued that every human life has equal value, and that no person or group has the right to make decisions concerning the lives of others on assumptions of inequality. It is not, for instance, justifiable to allocate scarce life-saving medical treatments on the basis of quality-of-life measurements.[6] On the other hand, even assuming that some human lives are more valuable than others, the judgments concerning them may require superhuman capacities. The traditional theological assumption is that while human beings are imperfect and their knowledge limited, God is omniscient. This implies that even if God, as an omniscient being, could pass valid judgments concerning human lives, the comparisons made by human beings would still be mere arrogant instances of playing God.[7]

As Chadwick herself notes, the playing-God objection may in this third form have some moral relevance as a reminder of the limits of our knowledge. It may also serve as a counsel against employing irrelevant criteria, like quality of life, in the inescapable human decisions concerning life and death. But the objection is not by itself sufficiently strong to refute any actual practices.

With regard to the technology aspect of the playing-God objection, Chadwick argues that divine omnipotence rather than divine omniscience provides the key to this side of the issue.[8] People who oppose activities like genetic engineering or artificial reproduction typically see these technologies as attempts to rival God's power by trying to create life or life forms.[9] When it comes to artificial insemination and *in vitro* fertilization, the counter-argument can be put forward, as in fact it is by Chadwick, that reproductive technology only aims at rearranging materials, not at creating previously non-existent entities. The same is not, however, quite true with regard to genetic engineering, which may, after all, create completely new life forms. Admittedly, new life forms have been created for centuries by animal and plant breeding. But these processes have been relatively slow, and humans have not been explicitly included in the programme. The opponents of genetic engineering may wish to argue that there are certain limits beyond which human beings cannot go without unlawfully playing God.

If this idea of fixed moral limits is taken seriously, the next step is to find out where the lines have been drawn and by whom. Chadwick considers three possibilities.[10] First, playing God can be understood literally, as a transgression of the invisible boundaries that separate immortal gods from mortal human beings. People who try to assume the role of gods are guilty of what the ancient Greeks used to call 'hubris', that is, of excessive pride. In the Greek mythology, overstepping the limits set by a divine will was generally punished in unusual and cruel ways. This literal interpretation of the playing-God objection is clear and intelligible, but its value as a moral guide is suspect. No critical morality can be based on the assumption that divine beings have set us limits which they continuously protect. Even if one believed in the existence of such divinities and in the sacredness of their will, it would be impossible to discover what the chosen deity would want us to do. In fact, one could well argue that the humans who pretend to be acquainted with the divine will are in fact putting themselves in the divine role, and thereby themselves playing God.

Second, the playing-God objection in the context of medical technologies may also be meant to state that the natural environment as a whole sets certain limits to our action. Humankind has during the last few decades acquired powers which could be used to destroy most of the biosphere. Many people seem to think that genetic engineering is one of these powers, and they fear that, for instance, the release of genetically altered organisms into the environment may have irreversible ecological consequences. Assuming that we are interested in the preservation of the biosphere, this objection against genetic engineering does indeed have some moral relevance. But the problem is that the appeal to consequences, which gives this argument its weight, also deprives it of its categorical disguise. It would, no doubt, be pragmatically unwise to destroy the only environment where we can live at present, but this does not amount to a categorical rejection of genetic engineering. The wrongness of the activity remains conditional upon the actual consequences.

Third, the limits of playing God can be set by human beings on the ground that certain actions, especially technology-related actions which have never been taken before, are liable to produce unforeseen, unpleasant and unpredictable consequences. Despite the appeal to consequences, this approach may be genuinely categorical, since no weight is given to the nature of the feared outcome or to the probability or improbability of its occurrence. According to Chadwick, the logic of the playing-God objection here is that the unknown consequences of going beyond (present) human limits cause fear, anxiety and uneasiness in many people.[11] Some of these people believe that an unimaginable disaster will meet us if new technologies are implemented. Others may have the feeling, unjustified, perhaps, but nonetheless painful, that divine retribution will follow the alleged human arrogance. And still others may be worried about the preservation of the current worldview, which may suffer from the confusion of its customary limits.

None of these negative feelings amounts, by itself, to an independent refutation of new technologies. But as Chadwick points out, the appeal to unforeseen consequences may be taken as advising us to be very careful in assessing certain delicate decisions. If the pros and cons of a given new technology are otherwise equal, the scales can be tipped by the unpleasantness inflicted on people by the mere thought of the innovation.

ARE GENETIC ENGINEERS PLAYING GOD?

Let me now summarize those parts of Chadwick's account which are relevant to my own question concerning genetic engineering and the alleged categorical wrongness of playing God. The development of gene technology is obviously subject to the remarks concerning unpredictable consequences. But, as we have seen, the only way in which this unpredictability can be brought to bear on the moral assessment of genetic engineering is through the fears and anxieties that people may have. This means that the technology aspect of the playing-God objection must be expressed in pragmatic and conditional terms, after all. Gene splicing, as an instance of 'playing God', is morally objectionable only if people's feelings are strong enough to outweigh the expected net utility of employing the technique. The playing-God complaint, understood in this way, is not a categorical claim, but an appeal to empirical facts which can be verified or falsified by observation and testing.

As regards the decision aspect of the matter, the playing-God objection can be interpreted as a warning against large-scale eugenic programmes. What I have in mind are genetic programmes which would aim at altering the human phenotype, either nationally or globally, to accord with the aesthetic or ethical views of scientists, politicians or others who claim to possess expert knowledge concerning the 'ideal human nature'. The possession of such knowledge would indeed require divine omniscience, and the majority of people would probably like to state that in this particular context the playing-God objection is valid without reservations. It should be noted, however, that the argument does not refute genetic engineering as such, but rather its misapplication to political purposes.

One does not condemn conventional medicine on the ground that medical skills can be employed in the execution of cruel and mutilating punishments. Similarly, one should not condemn biotechnology because it could be misused to create monstrous future dystopias.

Whether or not the argument against eugenic programming is categorical is a matter of some dispute. The theorist who prefers pragmatic interpretations can argue that the ultimate reason for rejecting the programmes is the decrease of happiness which is expected to result from them. The opponents of this view, in their turn, can reply that the expected outcome in terms of human pleasure or happiness is not decisive, since there are stronger, deontological reasons for banning designs to alter people. These deontological reasons cannot include appeals to 'playing God' or 'God's will', because those are the concepts we are trying to analyse here. But an alternative can be found in the claim that genetic engineering, especially when it takes the form of manufacturing human beings, is 'unnatural' or 'against nature'.

IS GENETIC ENGINEERING UNNATURAL?

The mere statement that an action is unnatural does not, of course, prove that the action in question is immoral, let alone that it ought to be banned.[12] Phenomena and practices which are rare, new, alien, or in any other way deviate from the everyday experience are often labelled by common sense as unnatural. But there is surely nothing inherently immoral in actions which are infrequent or previously not seen or heard of. More sophisticated analyses for the concepts of natural and unnatural are not always helpful, either. Take, for instance, the interpretations criticized by Ruth Chadwick: actions are unnatural if they interfere with the natural course of events, or put life on earth in jeopardy.[13] As for interfering with the course of nature, people have been doing exactly that for centuries, and most of the interventions have never been regarded as immoral. On the contrary, it could well be argued that once the significance of, say, hygiene, adequate nutrition and health care was discovered, it became a moral duty to employ these measures against the 'natural course of events', which would lead to diseases and starvation. And as for protecting life on earth, the reasons for a preservationist policy seem to be conditional rather than categorical. Whether or not an activity can be considered unnatural depends on its consequences, not on its intrinsic qualities. No logical connection exists between actions which are unnatural in the defined two senses and actions which are 'categorically' immoral.

A recent attempt to formulate and employ the argument of unnaturalness against genetic engineering can be found in the report of the Enquete Commission to the German Bundestag.[14] The Commission tackles three questions which are focal to the issue, namely the definition of the natural as opposed to the unnatural, the reasons for preferring naturalness to unnaturalness, and the division of different kinds of biotechnology according to their natural and unnatural characteristics.

As for the question of definition, the development of individual human beings is regarded in the report as natural only if it is not determined by technical production or social recognition. Technological and social processes can, according to the Commission's view, produce only unnatural artefacts.

The value of promoting naturalness and avoiding artificial elements in practices which concern human development is in the report linked with the need to protect the humanity and dignity of human beings. Our humanity, so the Commission asserts, 'rests at its core on natural development', and our dignity 'is based essentially on the naturalness of our origins'.[15] If technological or social interventions are allowed, then the result is that people will be made by other people, and the Commission regards this possibility with extreme suspicion. Human beings whose existence and personal qualities depend on the planning or caprice of other human beings are not free persons in the full meaning of the term, and their lives lack the individual worth of naturally developed human lives. It is the untampered chance of nature that secures our independence from other people, our personal freedom and our individual worth as human beings.[16]

These considerations lead to the following normative views regarding different kinds of human genetic engineering. First, somatic cell therapies performed on foetuses, infants and adult human beings are, at the moment, justifiable as experimental treatments. Whether or not such treatments should be abandoned or condoned in the future remains to be judged by their practical success. But the humanity of individuals is not threatened by the use of genetic medicine when the individuals in question have already developed into the beings that they 'naturally' are. Second, the mapping of the human genome is legitimate as long as it is employed to diagnose the need for somatic cell therapies. The potential use of gene maps for other (eugenic) purposes is more controversial. Third, cloning and large-scale eugenic programmes must according to the report be banned as gross instances of manufacturing people. And fourth, if a strict interpretation is given to the Commission's ideas concerning naturalness, germ-line gene therapies must also be prohibited. All interventions in the germ-lines of individuals diminish, according to the foregoing argument, their independence, uniqueness and worth as human beings.[17]

The opinions among the Enquete Commission diverged, however, regarding the legitimacy of germ-line gene therapies. Only some members of the Commission upheld the strict interpretation of naturalness, while others advanced a more moderate view. The core of the latter, moderate interpretation is that the medical corrections of obvious defects are not unnatural, as they do 'not manufacture the human genome capriciously, but measure it against nature, that is, good health'.[18] Illness and suffering can be a part of a person's identity, but if they are prevented before the person even exists, there is no point in maintaining that her or his individuality is unlawfully changed or manipulated. The genetical treatment of early embryos is, so the moderate reading goes, directly comparable to any conventional treatment of foetuses and neonates who cannot give their consent to the procedures.[19]

The divergence of opinions within the Commission is, no doubt, an interesting detail for those who believe that the unnaturalness objection is tenable as such. There are, however, several good reasons for thinking that this is not the case.

First, the argument from unnaturalness seems to apply to many practices which have been traditionally considered quite acceptable. If genetic engineering is to be condemned because of its power to change individuals by technical means, then most medical interventions should be condemned as well. Surgical operations, for instance, often alter people by transforming them from fatally ill patients into perfectly healthy citizens. And changes of personal identity may be even more drastic in the case of radical psychiatric treatments.[20]

Second, the Commission's argument presupposes theoretical elements which are by no means universally accepted. The report's entirely biological view concerning personal identity is a case in point. According to that view, human beings are who they are and what they are almost exclusively through the arrangement of their genes. Culture, education and social environment cannot significantly change the individual's identity: only biotechnology can do that. Very few philosophers today believe that such a strict biological definition of personality and individuality could be credibly defended.[21] Another presupposition in the report which can be criticized is its underlying view of human freedom and independence. The argument requires that human beings can be free from each other's influence in the sense that people are not 'manufactured' by other people. This is obviously true if the manufacturing of people is understood literally: human beings cannot at the moment be mechanically created by each other except in science fiction. But when it comes to less obtrusive types of interaction, it is also true that people simply cannot survive and function without the often restrictive and moulding presence of other people. Human freedom without the individual's dependence on others is only an abstraction with no reality to it.

Third, the unnaturalness objection presented in the report rests on the assumption that genetic engineering would undermine the worth, humanity and dignity of the individuals produced by the technique. This assumption is not only dubious, but it may be positively insulting towards those human beings who will be born in the future genetically altered or cloned, perhaps against prevailing laws. The depth of the actual insult depends upon the interpretation that one gives to the Commission's view. One possibility is to state that, according to the report, genetically engineered individuals would in fact lack humanity, dignity and personal freedom because their chromosomes had been tampered with. This line of argument would obviously be unreasonably unfair towards the individuals in question. Another possibility would be to assume that the Commission does not discuss the objective worth of human life in the first place, but the individual's subjective sense of worth in her or his life. The argument would then be that genetic engineering is wrong because the knowledge of one's 'artificial' and 'unnatural' origin reduces one's sense of worth and dignity. And yet another possibility is to claim that other people's adverse attitudes will make genetically engineered individuals unhappy.

The statements concerning attitudes can, no doubt, materialize under predictable circumstances. But since people's attitudes towards themselves and towards others are subject to change, the argument in this form is conditional rather than categorical. If genetically altered human beings can be expected to have difficulties in coping with the question of their origins, these difficulties may constitute a weak prima facie case against germ-line gene therapy, cloning and eugenic programmes. But this does not imply that these practices could be categorically rejected.

CONCLUSIONS

It seems, then, that appeals to the unnaturalness of biotechnology do not amount to tenable categorical arguments against its use. Nor do these appeals lend any support to the playing-God objections discussed in the above. In fact, the only thing of any normative value to be deduced from the unnaturalness objection is the fear that people may come to see and treat genetically engineered individuals as inferior to other human beings. But the significance of this point is minimal, since, first, people do not need to know who is genetically altered, and second, there is no reason to believe that people's attitudes towards each other would be dependent upon what they knew about each other's genomes.

On a more general level, it seems that the allegedly categorical objections against gene manipulation do not in the end merit the attention that they have been given in the literature. Such objections are employed frequently and without discrimination, but closer scrutiny either dissolves them entirely or reveals that their sound core, if any, is conditional or pragmatic. This is true about the playing-God objection, which can be interpreted as a counsel against forgetting improbable consequences and people's feelings. The same is true about the unnaturalness objection, which can serve as a warning against making people whose human worth and dignity will be questioned either by themselves or by others. But no decisive arguments for or against genetic engineering can be found from these quarters. The ultimate justification or rejection of biotechnology must be based on pragmatic considerations.

NOTES

I should like to thank Mark Shackleton, Lecturer in English, University of Helsinki, for improving the style of this paper.

1 The abbreviation 'DNA' stands for 'deoxyribonucleic acid'. The first successful human somatic cell therapy was reported by Timothy J. Ley and his colleagues in December 1983 in the *New England Journal of Medicine*. As for germ-line gene therapy, successful experiments have

been made since 1982 with (at least) fruit flies and mice, and it seems that there are no overpowering technical hindrances to medical success in humans. See, e.g., A. J. Varga, '"Playing God": the ethics of biotechnical intervention', *Thought* 60 (1985): 181–95; 'A report from Germany – an extract from *Prospects and Risks of Gene Technology: The Report of the Enquete Commission to the Bundestag of the Federal Republic of Germany*', *Bioethics* 2 (1988): 256–63; S. Kingman, 'Buried treasure in human genes', *Bioethics News* 9 (1990), no. 2: 10–15.

2 I forgo here the question whether or not these objections are well taken. The question is taken up by Heta Häyry in her contribution above.

3 R. F. Chadwick, 'Playing God', *Bioethics News* 9 (1990), no. 4: 38–46 (first appeared in *Cogito*, autumn 1989).

4 ibid., 40–2.

5 Cf. J. Harris, *Violence and Responsibility* (London: Routledge & Kegan Paul, 1980), 71.

6 Chadwick, op. cit., 41. Cf. J. Glover, *Causing Death and Saving Lives* (Harmondsworth, Middlesex: Penguin Books, 1977), 102–3. Glover seems to present two arguments concerning equality, but he does not differentiate between them. The first argument refers to the decision-maker's equality with those whose lives the decisions influence. Frankly, I do not see the point of this remark. My fundamental equality with other people cannot, as far as I can see, change the fact that if I am put in a position where decisions between other people's lives must be made, I must make them. The second argument is more intelligible to me, and refers to the mutual equality of those whose lives my decisions concern. This is the argument that Chadwick mostly employs, and the one that I have presented in the main text. See also J. Glover, *What Sort of People Should There Be?* (Harmondsworth, Middlesex: Penguin Books, 1984), 45–7.

7 Chadwick, op. cit., 42.

8 ibid., 43–5.

9 For a theological refutation of this objection, see Varga 1985, 186–7.

10 Chadwick 1990, 43–5.

11 ibid., 45.

12 On the (ir)relevance of naturalness to morality see, e.g., J. Radcliffe Richards, *The Sceptical Feminist: A Philosophical Enquiry* (Harmondsworth, Middlesex: Pelican Books, 1982), pp. 65 ff.; P. Singer and D. Wells, *The Reproduction Revolution: New Ways of Making Babies* (Oxford: Oxford University Press, 1984), pp. 36ff.; J. Harris, *The Value of Life: An Introduction to Medical Ethics* (London: Routledge & Kegan Paul, 1985), pp. 186ff.

13 Chadwick, op. cit., 40, 44.

14 'A report from Germany', op. cit., 257–8.

15 ibid., 257. The whole of the quoted passage is as follows: 'The starting-point of every evaluation must be that the humanity of human beings rests at its core on natural development, not on technical production and not on a social act of recognition. The dignity of human beings is based essentially on their being born and on the naturalness of their origins, which all humans share with each other.'

16 ibid., 257. In full: 'The fact that human beings are not the project and the planned experiment of their parents, but are the product of the chance of nature, secures the independence of human beings from each other, their individual worth . . . "to make the formation of our genotype . . . dependent on the caprice of other people is incompatible with the essence of a free person".' The latter passage is quoted in the report from R. Loew, 'Gentechnologie: vom Können zum Dürfen – philosophische Überlegungen', *Die Neue Ordnung* 38 (1984): 176ff.

17 ibid., 258.

18 ibid.

19 There are two intrinsic problems within the 'moderate' view, which are not essential to my main line of argument, but which are nevertheless worth mentioning. First, defining 'good health' is not an unambiguous matter, and one could well argue that human caprice always enters germ-line gene therapies through the particular definition employed. Second, the moderates of the Commission seem to assert that eugenic programming can be categorically prohibited because of unnaturalness even though germ-line gene therapies cannot. This is a highly controversial view which presupposes that a tenable distinction can be drawn between the two practices.

The alleged difference between gene therapies and genetic improvement programmes is that the former are aimed at eliminating hereditary diseases while the latter are intended to bring about or intensify some positive qualities in future individuals. This distinction is unclear, as it is obvious that illness may hinder the development of certain positive qualities and promote the development of others. A physically disabling disease, for instance, may prevent the individual from being strong and athletic (which are often regarded as 'good qualities'), but it may indirectly promote the individual's willingness to learn useful cognitive and artistic skills (which are also often considered good).

The main point to be noted here is, however, that the argument which supports gene therapy on individuals can also be used to defend demographic genetic changes which are not, strictly speaking, disease-related. The justification for bypassing the patient's consent is, according to the moderate view, the best interest of the patient her or himself. How could positive improvements in the patient's best interest be ruled out if the elimination of negative factors in the patient's best interest is accepted?

20 Besides, as the Commission's moderates note, there are social arrangements, such as marriage rules, which have been interfering with the 'natural' human gene pool for millennia, thereby modifying the qualities of individuals. To condemn germ-line gene therapies categorically would require the condemnation of all these social practices as well. See 'A report from Germany', op. cit., 259.

21 For a critique see, e.g., H.-M. Sass, 'A critique of the Enquete Commission's report', *Bioethics* 2 (1988): 264–75, 269.

12

BIOTECHNOLOGY, FRIEND OR FOE? ETHICS AND CONTROLS

John Harris

The ethical and policy implications of developments in biomedical science are increasingly dominating not only public debate but also scientific meetings and international fora. There is increasing recognition of the necessity for developing a moral framework that can keep pace with and control rapid change. In many different areas of biomedical science, technological advance has posed new dilemmas and necessitated the reassessment of our moral categories.

The evidence of our efforts to respond to these challenges is pervasive. On the one hand a host of dramatic legal cases throughout the world have brought to public attention not only the unprecedented powers of the new science but also the human problems it brings with it, both positive and negative. Spectacular biomedical success stories such as the birth of test-tube babies or the announcement of the discovery of the genes responsible for specific genetic disorders have sent governments and jurisdictions scrambling for basic principles which might illuminate the management of the legal, practical and human problems which inevitably follow. It is the area of biotechnological advance that has recently most sharply focused the issues. World attention has been drawn to the need to make commensurate advances in our moral assessment of scientific progress and in our ability to anticipate developments and not simply to respond to them.

In this chapter[1] I wish to examine two recent and influential attempts in the United Kingdom to respond to these needs, and by studying their deficiencies, to highlight some of the problems which attend one particular sort of response to the challenge of formulating methods of controlling biotechnological advance.

In the United Kingdom there is a strong tradition of addressing problems such as we have described by establishing a government committee to examine the issue and make recommendations for the framing of legislation. Contemporary examples include two committees under the chairmanship of the philosopher Dame Mary Warnock, one to address the problems of people with disability[2] and one more recently to examine human fertilization and embryology. The most recent of such government committees is that chaired by Sir Cecil Clothier – the Committee on the Ethics of Gene Therapy,[3] which like its predecessors is now usually referred to by the name of its chairman, in this case as 'the Clothier Committee'.

In parallel with attempts by government to address the issues it has been common for other interested non-governmental bodies to establish similar committees with a view to formulating policy and influencing legislation. The most recent of these, addressing the same issues as Clothier but rather more fully, was a working party of the British Medical Association which published its report on 'The science and ethics of genetic technology', entitled *Our Genetic Future*[4] in the late spring of 1992.

My present interest is in the similarity between the conclusions of the two independent reports and in similarities between the deficiencies of these conclusions and in the reasoning which generated them. I will look very briefly at the reports in turn and then attempt to draw some general conclusions.

I CLOTHIER

There are three separate and revealing problems exhibited by Clothier's approach to the question of the ethics of gene therapy. The first has to do with the ethical foundations of the study, the second with its apparent account of what constitutes disease, and the third with its failure adequately to consider germ-line therapy.

1 The ethical basis

The ethical basis of the Clothier Report is sketched in thirteen short paragraphs. I would wish to take issue with every one of them, but I will confine my observations to three.

The first paragraph of this section of the report outlines the starting point:

> We begin from the basis that ethics are the moral convictions of thoughtful, conscientious and informed people, to be rejected only when they are in conflict with other convictions which stand better the test of reflection.[5]

There are all sorts of problems here. Do ill-informed people have moral convictions, can these convictions be ethical and if so, how? An underlying problem is that of giving an account of how a conviction differs from a prejudice, however thoughtful and informed the people holding it may be. Clothier's answer to this question seems to reside in the idea of 'the test of reflection'. Such a test, if it is to do the job of identifying sustainable moral convictions, would at the very least have to involve the subjection of moral convictions to critical examination; in particular to examination of the consequences of holding them, of their consistency with other principles and beliefs, and of the evidence and argument which might support and sustain them. With this thought in mind we must see whether the remaining account of the ethical basis of Clothier might meet such a test. The task is: 'To find an ethical position which would command acceptance in this country for the foreseeable future'. To this end Clothier reports:

> we have consulted widely among individuals and bodies representing diverse religious and cultural beliefs. We have also taken notice of the treatment of these matters in other countries of similar cultural heritage.[6]

Now all his sounds fair enough, and indeed it is, until one notices how such consultations figure in the development of a position. Like its predecessor the second Warnock Report, Clothier sees its task not as leading and guiding public opinion but in following and conforming to it; not in subjecting the moral convictions it finds to the critical examination that one would expect members of such a committee to be pre-eminently able to provide, but seemingly in finding some principles which appear to be relevant to the task in hand and which the members believe are widely accepted in the community. These are the two principles which commended themselves to Clothier:

> The first ethical principle which has commanded our attention is the obligation inherent in human nature to enquire, to study, to pursue and apply research by ethical means.[7]

> The second ethical principle which has governed the deliberations of the Committee is that in the sometimes inescapable tensions between the pursuit of knowledge and the protection of patient's interests, the latter must prevail.[8]

From a committee which has examined the science of genetics the first principle is certainly audacious. In which portion of the human genome Clothier located the gene or genes for 'the obligation to enquire, to study, to pursue and apply research by ethical means' we are not told, but since on the highest authority it is 'inherent in human nature', we may expect an imminent announcement.

This principle, in all its bizarre complexity, has the air of something constructed to fit the conclusions the committee wished to arrive at rather than something so compelling, obvious and generally accepted that its primacy was inescapable.

The second principle seems equally unreflective, for the pursuit of knowledge is necessary to protect patients' interests both now and in the future. The principle implies that existing patients' immediate interests should prevail over the future interests of existing patients *and* over the interests of future patients, even though these future patients will be numerically more significant. While such a principle might ultimately prove defensible and even attractive, it cannot claim to be one that thoughtful, conscientious and informed people ought immediately to accept without being offered reasons and arguments for so doing.

2 The Nature of Disease

In concluding its guarded approval for gene therapy Clothier states at paragraph 4.22 that:

> We are firmly of the opinion that gene therapy should at present be directed to alleviating disease in individual patients, although wider applications may soon call for attention (see 2.15). In the current state of knowledge it would not be acceptable to use gene modification to attempt to change human traits not associated with disease.

Now this again sounds responsible and plausible, but its plausibility depends upon not pressing the question of what constitutes disease or what exactly is a trait associated with disease in individual patients.

This is clear from the fact that this passage refers us back to paragraph 2.15 for a gloss on the idea of wider applications, that is, applications that go beyond those directed to alleviating disease in *individual patients*. Paragraph 2.15 states:

> Gene therapy has wider possibilities for medical practice than the correction of single gene disorders. For example, it is being investigated as a possible new approach to the management of a wide spectrum of diseases, ranging from infections such as AIDS to cancer, and it is being studied as a means of strengthening the body's immune response to viral infections.[9]

Clothier would appear to rule out all of the applications contained in 2.15, since it specifically classifies them as 'wider applications' and distinguishes them from therapy directed to alleviating disease in individual patients.

The only justification offered for denying the human community the benefits described in 2.15 is that 'In the current state of knowledge it would not be acceptable . . . to attempt to change traits not associated with disease'. But aside from the equivocation between 'traits associated with disease' and 'disease in individual patients' the principle is prima facie implausible and certainly unpersuasive without supporting argument. We are surely entitled to know why it would be unethical to attempt to manage diseases like AIDS or cancers using gene therapy.

However, the arbitrariness and indeed spuriousness of the distinction between traits associated with disease and traits not so associated can be seen if we turn to wider applications still, applications which would be ruled out by Clothier even if the rejection of what the report *calls* 'wider applications' is simply the result of poor drafting and unintended. Take a trait like intelligence. Most people and certainly Clothier would not regard intelligence as a trait associated with disease at all, let alone disease in individual patients. Clothier, I believe, firmly intends to rule out the modification of traits like intelligence or, say, physical height or build. Now, as it happens, intelligence, multifactorial and ill understood as it is, holds no immediate prospect of genetic modification. But it is interesting because of the way it highlights the possible consequences of Clothier's opinions here.

Imagine two groups of mentally handicapped children, or children with severe learning difficulties. In one the disability is traceable to a specific disease state or injury. In the other it has no apparent cause. Suppose now that gene therapy held out the prospect of improving the intelligence of children in both groups so that they could lead normal lives.[10] Clothier seems to say that it would unethical to attempt to help either group, but might possibly be interpreted to mean that we could help those whose condition is traceable to a disease but not the remainder. The same arguments might be applied to some forms of dwarfism, or restricted growth, and hence to possible modifications in height, build, and other traits not associated with disease. How either of Clothier's conclusions could be ethically defended here requires more conscientious thought than I am able to command.

Finally Clothier dismisses germ-line modification in two short paragraphs. I will note these now but deal with the problem of objections to germ-line therapy in more detail when I discuss the second report, *Our Genetic Future*, in a moment. Clothier has two main grounds for the dismissal of germ-line therapy. The first is that 'there is at present insufficient knowledge to evaluate the risks to future generations' and the second that it is in any event unnecessary because 'embryonic diagnosis and selective implantation of an unaffected embryo would provide another way of achieving the same end without incurring unknown risks.'[11]

II OUR GENETIC FUTURE

The report from the British Medical Association's Professional, Scientific and International Affairs Division called *Our Genetic Future*, published as a book, is a much fuller and more wide-ranging document than that generated by HM Government. However, even the BMA's committee members cannot resist the temptation simply to parade their prejudices on occasion. 'The possibility that some day parents may be able to "order" children with particular characteristics such as the ability to do mathematics is not only abhorrent but also unlikely ever to be achieved.'[12] That this is abhorrent is treated as beyond doubt; why it ought to be abhorrent we are not told.

For my part I do not find it abhorrent, and I wonder why, if it is legitimate to try to *educate* our children to acquire the ability to do mathematics, it is not legitimate genetically to engineer into them a like ability. Why, ethically, might it be wrong to attempt something via genetic engineering that it is not wrong to attempt via what used to be called 'social engineering'? There may, of course, be a good reason, but in the absence of any attempt to provide one, what we are offered is insight into the personal aversions of the committee members and contributing authors. The disturbing feature of this excursion into gut reaction from a distinguished committee of an equally distinguished organization is the way that it seems to take pride in its prejudice. The BMA would presumably feel the same abhorrence at the thought of engineering musical ability, or rather the potential for it, into children, and yet many parents start rigorous musical education at a very early age with the same object in view.[13]

If it were not the case that this prejudice is clearly widely shared, one might detect here a typically British distaste for efficiency. The underlying explanation might be that gene therapy, because it is so much more efficient than education, so much more likely to constitute an effective method of imparting mathematical ability, is abhorrent primarily because it leaves no room for human error. However, this would surely involve a dubious view of the mechanism by which a gene or genes for mathematical ability might operate. They would surely only confer a predisposition to mathematical ability which might well be thwarted by poor education or lack of motivation on the part of the subject. Whether the inefficiency of both gene therapy and education is a cause for comfort or celebration is, of course, a further and separate question.

Germ-line versus somatic-line therapy

Both Clothier and *Our Genetic Future* treat germ-line and somatic-line therapy as if they were alternatives between which we might choose on each occasion depending on the merits of the different routes to the same objective. And the suggestion underlying such an approach is, of course, that if these are genuine alternatives, the less problematic should be preferred. However, it seems likely that there will be some (perhaps many) conditions for which germ-line therapy offers the best or maybe the only prospect of cure, and of course there are some parents, those who can have only genetically damaged children, for whom germ-line therapy will offer the only possibility of producing unaffected children.[14] We would need weighty reasons indeed to deny therapy to those for whom operating on the germ-line offers the only prospect of help.

I have examined most of the standard objections to germ-line therapy elsewhere and found them wanting.[15] I will not repeat most of those objections here, but I will just deal with a few which are prominent in *Our Genetic Future* and which include the two mentioned by Clothier.

Clothier simply mentions the risk to future generations, but *Our Genetic Future* provides some detail. A germ-line error, the BMA tells us:

> may not even manifest itself during the lifetime of that person, if it is a recessive mutation, but only appear years hence as a homozygous trait in the descendants. By that time the possibility of correcting the mutation may be long past and the affected individuals and society will be required to live with the consequences. Until gene therapy can be shown to be absolutely accurate in inserting and targeting new genes, the potential hazards of genetic modification in germ cells are too grave to permit its use.[16]

There are many questions and problems in connection with this passage. Why is it assumed that the error will prove impossible to correct? Is the test of 'absolute accuracy' suggested for germ-line therapy more stringent than the test required for somatic-line therapy, and if so, why? One suspects that it is and that the reasons are given in the passage. However, it is instructive to compare the BMA's attitude to the future dangers to individuals and to society resulting from germ-line therapy with those resulting from failure or inability to treat genetic disorders. Just a few pages later, discussing pre-natal diagnosis, we find:

> There are several options open to couples known to be at risk of having a child affected with a genetic disorder. They may decide not to have their fetus tested and have the baby regardless. Alternatively they may choose to have the test and on the basis of the result decide whether to seek a termination of pregnancy or not.[17]

These alternatives are presented without comment or stricture by the BMA, as they usually are elsewhere, as matters entirely within the unfettered discretion of parents. Where are the dire warnings that it may prove impossible to correct the problem and that 'society will be required to live with the consequences'? Why are not these consequences likewise 'too grave' to be permitted? If, to help these parents have the children they want, it is justifiable to impose on subsequent generations and society the *certainty* of such grave burdens, why is it not legitimate to *risk* such burdens as an outside chance of things going wrong in germ-line therapy?[18]

2 Research and redundancy

Two further objections which recommended themselves to the BMA require consideration. The first is the claim that since germ-line therapy requires research on embryos and since 'Research on embryos gives rise to highly emotional debate, the moral status of the early embryo remains disputed. Consensus is unlikely ever to be reached on this particular issue as each camp – the "pro-life" and the pro-research lobbies remain steadfast in their arguments.'[19] But as the BMA go on to admit, research on embryos up to fourteen days is permitted in the United Kingdom, and such objections have not proved decisive against the use of *in vitro* fertilization techniques, which also depend on embryo experimentation.

Finally there is 'a very good practical reason for opposing it. It is unnecessary; simpler techniques can be used to prevent a disease-causing gene from being inherited, that is by preimplantation diagnosis'.[20]

There are important ways in which this claim is at best misleading and at worst simply false. In the first place it is plausible only on the assumption that couples at risk will reproduce only via IVF and will accept the implantation only of healthy embryos. Some people who want children might well prefer children with genetic disease to no children at all if that is the only option for them. Others, we already know, decline IVF and other screening and reproduce hoping for healthy children. These same parents might, if germ-line therapy were available, reproduce only via IVF,[21] availing themselves of the chance germ-line therapy offers to have unaffected children.

As we have already noted, *Our Genetic Future* itself admits these facts without comment or criticism. In the light of this we know that a ban on germ-line therapy would mean that children with defects in the germ line (many of which would result in catastrophic disease) would continue to be born with the terrible effects on themselves, society and their parents noted by both reports. This would certainly be the consequence of the policy recommended by both Clothier and the BMA. Conversely, permitting germ-line therapy[22] might have no such effects, although, and of course, some children with defects in the germ-line will continue to be born whatever we do. At worst germ-line therapy would only create the *chance* of unforeseen deleterious consequences rather than the certainty of them so readily embraced by both reports.

Taking stock, we can say that the radical rejection of germ-line therapy found in Clothier, in *Our Genetic Future* and elsewhere is misleading if it implies that persistent damage to the germ-line of many humans can be avoided if germ-line therapy is banned and will be the inevitable or likely consequence of permitting germ-line therapy. On the contrary, we know that some damage to the germ line of many humans will be prevented if germ-line therapy is perfected and permitted. We do not know that any new damage will occur, and if it does, it may prove remediable. Of course we will need to be pretty confident that we can target gene insertions into harmless sections of the germ-line, as we will with somatic-line therapy.

In some circumstances, different standards for the accuracy of insertions and safety of the procedures are appropriate as between somatic-line and germ-line therapy. If the subject of somatic-line therapy can choose for himself whether to undergo certain risks in the hope of possible benefits, then I believe this is a matter for his unfettered discretion.[23] Where somatic-line therapy on a young child or embryo is contemplated, then, as with germ-line therapy, we must be very confident of two things: first, that the procedure has a high chance of success,[24] and second, that such risks as remain are worth running given the severity of the alternative of not running them. The standards set for the accuracy of permissible germ-line therapy must not, however, be too severe. We already accept the risks of radiation and chemotherapy, which affect the germ-line and do not always result in sterility. A total ban on germ-line therapy might pre-empt the research which would make perfection of the technique possible.

III CONCLUSION

In attempting to demonstrate the shortcomings of two of the most recent attempts by committees to wrestle with problems in bioethics, I have not, of course, shown that such committees are necessarily flawed or are more error-prone than the rest of us. Nor can I claim to see clearly what mechanisms we should use as the basis for the formulation of social policy and the framing of legislation. I want to conclude by making a plea for using such committees in a particular way and especially for declining, except in the case of the clearest and most urgent necessity, to legislate in these sensitive and controversial areas.

Committees charged with such heavy responsibilities too often see their task, as did Clothier, as finding 'an ethical position which would command acceptance in this country for the foreseeable future'. They tend to believe that the best way of achieving this is to arrive at a position which already commands wide acceptance. This means that they almost inevitably follow popular prejudice, the views of those who have come to swift judgments without the benefit of the expertise, access to evidence and information and time for reflection available to the committee members. I believe that there is no point in commissioning committees of experts in biomedical science and ethics if they constrain themselves to simply follow the opinions of people who are not so expert and have not had access to the same information. Their task, I believe, is to attempt to lead public opinion, not to follow it.

Equally, once a major committee is established to look at the ethics of any procedure, whether by the government or by a major national or international body like the BMA or indeed the European Commission, there is a tendency to draw a dangerous conclusion and to follow it with an equally dangerous tendency. The dangerous conclusion is that the ethics of whatever is under consideration have now been, or are being, conclusively dealt with and there is no further need for worry on this score. The dangerous tendency is to use the findings of such a committee, however widely, effectively and conclusively criticized, as the basis for legislation. This happened in the United Kingdom following the Warnock Report on human fertilization and embryology, the findings of which were largely accepted as the basis for legislation despite widespread criticism and despite the fact that the committee members published three separate expressions of dissent.

Even where such committees themselves invite responses and criticism, and even where governments intend only to use such committees to inform decision-making rather than dictate it, there is a great tendency to legislate on the basis of committee findings. This is for the obvious and natural reason that once a committee has been selected, been constituted, heard evidence and reported, there is no more authoritative source of advice available. I suppose it must just seem silly to have set up such a committee and then to ignore what it says; and so, except where political expediency dictates otherwise, such committees often provide the framework for legislation.

Of course it is easier to see problems than solutions, but equally we should not embrace solutions that do not solve our problems. In particular it is one thing to come to a view about the ethics of biotechnological advance, quite another to solidify such conclusions in legislation.

We need, I believe, to recognize two important features of ethical debate in the area of public policy. The first is that this debate is ongoing and open-ended. That ethical problems are never finally solved, though advances are possible. In the light of this we must welcome and accommodate continuing public debate of all ethical issues. The role of ethical committees is, I believe, to lead and inform this debate, not to pre-empt it or bring it to a close.

Second, we need to recognize and respect the fact that many different ethical judgments and conclusions are possible from decent people of goodwill. The abortion debate is a classic example, and it typifies debates on many issues in bioethics. The people on both sides of this debate are for the most part principled people who disagree profoundly about an issue of the highest importance. It is crucial that one side does not force its conclusions on the other but attempts to persuade rather than legislate. I cannot improve on Thomas H. Murray's analysis of the reasons for this:

> What would happen if a convinced minority[25] succeeded in imposing its will on the majority? Those individuals would have achieved their goal of discouraging officially a practice they believe is seriously immoral, and in that way they will have met their obligation to do what they believe is right. But by imposing their will on so many others, they will have threatened other moral values, perhaps equally dear. They will force others to pursue their vision of human good furtively. In all likelihood, abortions will continue to be done, though under more dangerous conditions; therefore, some women will be injured or killed. A number of good people will commit what will now be defined as criminal acts. Widespread disobedience to the law and resentment against what will be seen as the imposition of oppressive and unjustified rules will feed discontent and loosen the bonds of social solidarity. Disobedience of one law may lead to disrespect for the law in general. The price then of a social policy intended to bolster one moral value may be terribly high: nothing less than an attrition of other moral values, values crucial to sustaining social fabric.[26]

In the areas with which we have been concerned here, germ-line versus somatic-line therapy, for example, and in the ethics of gene therapy more generally, I do not believe the arguments are sufficiently strong or clear for some people to legislate away the possibility for others to achieve their aspirations for themselves and for their children. Nor, crucially, are they strong enough to justify preventing medical science developing in these areas. These aspirations will include the aspiration to have healthy children or children with other desired features, like gender, build, or even mathematical ability if this proves possible and safe.

Even if this point were to be conceded, it would be important to ensure that people's chances to pursue their own vision of the good in their own way were not legislated away by covert means. In particular, bearing the lessons of the present reports in mind, the standards of safety required should not be rigged so that they effectively rule out germ-line therapy.

NOTES

1 Some of these ideas were first presented at the *Assize Internationale di Bioetica*, Rome, 29–30 May 1992, at the Auletta dei Gruppi Parlamentari.
2 DES, *Special Educational Needs*, London: HMSO, 1978.
3 *Report of the Committee on the Ethics of Gene Therapy*, Presented to Parliament by Command of Her Majesty January 1992, London: HMSO.
4 The British Medical Association, *Our Genetic Future*, Oxford: Oxford University Press, 1992.
5 *Report of the Committee on the Ethics of Gene Therapy*, 3.1, p. 10.
6 ibid., 3.3, p. 10.
7 ibid., 3.4, p. 10.
8 ibid., 3.6, p. 10.
9 ibid., 2.15, p. 6.
10 A similar point was made tellingly by Dr Richard Newton at a Forum on the Ethics of Gene Therapy which took place at the Royal Manchester Children's Hospital on 14 April 1992.
11 *Report of the Committee on the Ethics of Gene Therapy*, 5.1 and 5.2, p. 18.
12 *Our Genetic Future*, chapter 13, p. 183.
13 Mozart was, of course, among many such children to 'benefit' from such an early start.
14 This will happen where a couple in which both partners have a recessive disease like cystic fibrosis want children.
15 In my *Wonderwoman and Superman: The Ethics of Human Biotechnology*, Oxford: Oxford University Press, 1992.
16 *Our Genetic Future*, chapter 13, p. 187.

17 ibid., p. 194.
18 It may be that a number of current therapies affect the germ line of patients, particularly radiation and chemotherapy in cancer treatment, and yet these are permitted. Of course men in particular are usually (though not necessarily) infertile after such treatment, though women may not be.
19 *Our Genetic Future*, p. 187.
20 ibid., p. 188.
21 Or having had both sets of gametes modified if this proves a viable method of germ-line therapy.
22 Once it had proved sufficiently reliable (as reliable as the somatic-line therapy would have to be to be approved by either Clothier or the BMA).
23 See the discussion of autonomy and decision-making in *The Value of Life: An Introduction to Medical Ethics*, London: Routledge, 1985, reprinted 1989, 1990 and 1992, chapter 10.
24 But not necessarily an extravagantly high chance of success. We should compare the current prognoses for, say, cancer therapy in children or attempting to correct spina bifida in children.
25 Of course the point would remain good if we were not talking of minorities.
26 Thomas H. Murray, 'So maybe it's wrong: should we do anything about it? Ethics & social policy', in William B. Weil Jr and Martin Benjamin (eds) *Ethical Issues at the Outset of Life*, Boston: Blackwell, 1987, p. 240.

13

GENETIC ENGINEERING AND ETHICS IN GERMANY

Ulla Wessels

1 WHAT THIS CHAPTER IS ALL ABOUT

The expression 'genetic engineering' refers to all procedures dealing with the artificial recombination of genetic material.

Here are some examples of the activities in question: in the field of environmental protection, the construction of micro-organisms capable of dismantling harmful materials; in the field of nutritional and agricultural sciences, the development of new strains of grains and domestic animals; and in the field of human medicine, efforts to develop genetic therapies and to produce gene-based medicaments, diagnostic aids and vaccines.

It is hardly surprising that the ethical problems involved (though, as we shall see, not their discussion) are the same in Germany as elsewhere: What are the risks inherent in genetic engineering? How can these risks be weighed against the advantages that genetic engineering promises? What consequences really *are* advantages? Should man be allowed to change 'human nature'? Does the alteration of genetic material damage the interests of those affected? And so on.

This chapter does not attempt to show how one *should* go about answering these questions. What we are asking is how these questions *are in fact* being dealt with in Germany. More explicitly:

I Which topics are being discussed most intensely?
II Who is primarily involved in these debates?
III Which beliefs and arguments play an important part?

Ad (I): The most controversial area of application is, unsurprisingly, man. It is this area, therefore, that we shall concentrate on in the following survey.

Ad (II): Those who, first and foremost, participate in the discussion of genetic engineering of human beings are lawyers, politicians, theologians, and a few philosophers. In addition, lobbies and institutions representing affected branches of industry and science as well as several independent pressure groups all publicize their views.

And now, for the rest of this chapter, ad (III). I shall, as far as this is compatible, try both to give a *structured* survey of the main types of arguments and to provide, in each case, numerous references enabling the more curious reader to locate the relevant sources himself or herself.

2 THE SEARCH FOR SAFETY STANDARDS

In February 1975, 140 scientists from 17 countries met in Asimolar to discuss DNA-replication experiments, the associated dangers, and possible safety measures. Just a few months later, in 1976, (West) Germany's Federal Ministry for Research and Technology[1] (henceforth: BMFT) set up an expert committee, composed entirely of scientists, on 'Safety Regulations for Research into the In Vitro Recombination of Nucleic Acids'.[2] In March 1977, the committee submitted the first proposal for safety regulations. They were practically identical to those proposed in the USA in 1975 by the NIH (National Institute of Health). On 21 January 1978, the German committee produced the *Regulations concerning the Prevention of Danger from the In vitro Recombination of Nucleic Acids.*[3] These regulations were modified over the next few years to cope with questions arising from new developments in research. However, they were binding only for federally funded research projects (at Max Planck Institutes, by the German Society for the Advancement of Scientific Research,[4] etc.) and for research projects under the responsibility of the federal states (mainly those at universities); they did not apply to private or industrial research.

As these regulations did not bind everybody, a *law* regulating the recombination of DNA was felt to be necessary. The first proposals in that respect were developed by the Enquete Commission 'Chances and Risks of Genetic Technology',[5] which was set up by the German Parliament in June 1984 and presented its final report in 1987: *The Report of the 10th German Parliament's Enquete Commission, 'Chances and Risks in Genetic Technology'*,[6] referred to, henceforth, as *BCRG*.

A first draft of the *Law to Regulate Questions of Genetic*

Technology (*GenTG*),[7] following, to a large extent, the *BCRG* suggestions, was passed by the Federal Cabinet in July 1989, and the law became effective on 1 July 1990.

The essentials of *GenTG* are these:

§ 5: a board of experts within the Federal Ministry of Youth, Family, Women and Health[8] (BMJFFG), called the Central Commission for Biological Safety,[9] is to be established; its task is to watch over the observance of *GenTG*;[10]

§ 6: a registration of all research in genetic technologies is made compulsory; exact records must be kept of the aims of research as well as all steps and procedures within the research process;[11]

§ 7: research in genetic technologies is classified on four safety levels;[12]

§ 8: research plans of the least risky types are subject to registration only; all others are subject to permission.

Right from the beginning, *GenTG* has not been without critics. While environmental organizations think that the safety regulations concerning the release into the environment of genetically altered organisms are too lax and that the public has too little control over such activities,[13] the industry feels that, by the compulsory disclosure of their plans and the partly tedious procedures of registration and permission, their foreign competitors will get the better of them.

Meanwhile, the companies' critique has triggered endeavours to modify *GenTG*; work in genetic technologies is likely to become less restricted. In particular, the national and international trade with genetically altered organisms and their release into the environment will probably be made easier; it would be permissible to register projects, and to request permission for them at shorter notice; and the right of the public to a say would be restricted. These modifications can be expected to become effective in 1993 or 1994.

3 THE SEARCH FOR MORAL STANDARDS

The problem of finding safety regulations is, roughly speaking, *how* to make sure, by technical or political means, that threats to life and health will not arise. That such threats *are* bad and *ought* to be avoided is a moral truth too obvious to be denied, and hence too obvious to require large-scale discussions.

But there are, of course, genuinely *moral* problems involved in genetic technologies. People wonder (or dispute about) what, in the end, we ought to do; what actions we ought to refrain from; which types of interference with nature and which results of such interference are desirable, and which are not.

These problems receive more attention the closer genetic technologies come to areas considered as morally sensitive, like (paradigmatically) man.

The issues here will be clarified as we go along, but for the moment, they can be classified as shown in Figure 13.1.

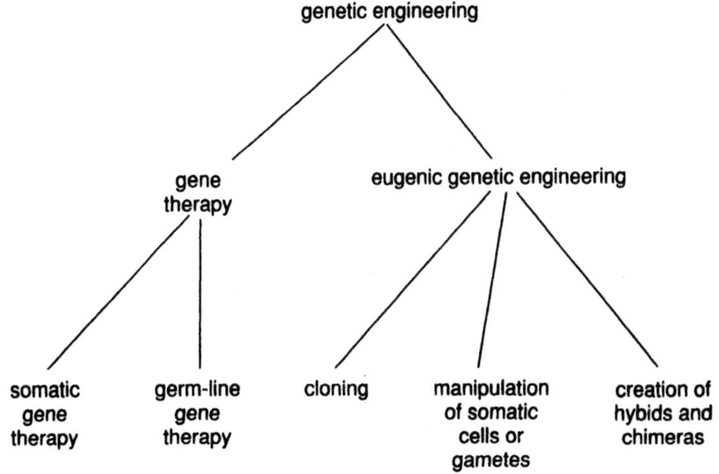

Figure 13.1 Genetic technologies

3.1 Gene Therapy

3.1.1 Somatic Cell Gene Therapy

Somatic cell gene therapy is the correction of genetic deficiencies in *somatic* (i.e. *body*) cells by production of new DNA and its insertion into these cells (see Suzuki *et al.* 1989: 727, 735). Hence, somatic gene therapy is only concerned with the full or partial healing of the *patient who actually exhibits the genetic defect*.

In Germany there is a widespread consensus that this is a good thing: as soon, that is, as the techniques for such treatment are actually perfected – which is considered likely to happen.

The argument here is based on the method's strong similarity to conventional and certainly unobjectionable medical procedures:

> Somatic gene therapy is a special form of substitution therapy. The absent body function is replaced not by hormone injection or the transfer of foreign organs but by transferring genetic material in the form of cloned DNA. The transfer of genes must be evaluated in the same way as the transfer of living material.
>
> (*BCRG*, p. 83)[14]

The comparability of somatic gene therapy with other medical methods also determines the legal position. Though very little is known about gene transfer into somatic cells, and though this practice, therefore, is viewed primarily as an experimental measure on human beings, it is possible to subsume it under the traditional legal concept of a 'new ground operation' (*Neulandoperation*). Thereby, the question whether *in principle* somatic cell gene therapy is legal is already answered in the affirmative; it is 'only' the exact conditions which are legally restricted. In particular, the patient's informed consent does not suffice to justify the interference with his 'physical integrity'; a *concrete* medical cost/benefit analysis is needed as well.[15] With this, the German legislation moves within the internationally accepted norms, and, specifically, in accordance with the guidelines laid down by the European Medical Research Council in 1987.

The boundary between illness and pure abnormality may be difficult to define. The main risk of somatic gene therapy, therefore, is seen in its possible use for breeding purposes (see below, section 3.2). For that reason it has been requested, even by those welcoming the *medical* possibilities of somatic gene therapy, that there be 'a clear catalogue of inherited illnesses that come into question for such treatment' (*BCRG*, p. 183).

Most of those who refuse to support even truly therapeutic somatic gene therapy and call for a clear prohibition of further research use varieties of arguments from (allegedly) false priorities.

The first argument is the argument against reductionist medicine. It goes without saying that the curing of illnesses is in the interest of mankind. Yet, for the Green Party, for example, somatic gene therapy is part of 'a tradition of medical activity that conceives of a human being as an ensemble of separate cells/organs treatable in themselves. The influence of such factors as the condition of the body as a whole, the history of the illness and the reasons for its development are thereby marginalised or even negated' (*BCRG*, p. 324). A policy aimed at the welfare of mankind and of the environment would consist not only in the healing or alleviation of maladies, but more importantly prophylactic measures.[16]

Foresight is certainly better than hindsight. But what if we *did* take all conceivable precautions and *then* fell ill (as is, no doubt, possible)? It is hard to see how an appraisal of prophylaxis can provide an argument against cures.

The second argument is the argument from injustice in the allocation of medical resources. Considerable means (money and research capacities) are being deployed for developing somatic gene therapy, although the number of illnesses treatable that way is small and their occurrence rare. Given that resources are limited and that four-fifths of the world's population have no access to modern medical care at all, the luxury of somatic gene therapy cannot be justified morally.[17]

Of course, this argument does not succeed in making a point against somatic cell gene therapy in general, but is, at most, a demand that other problems should be dealt with simultaneously or initially.

3.1.2 Germ-line Gene Therapy

While there is a widespread, though not total, consensus in Germany that somatic gene therapy can develop into a good thing, there is a comparable quasi-consensus *against* the removal of genetic deficiencies by inserting DNA into germ cells (eggs and sperm) or into the cells of a pre-embryo (i.e. cells that give rise to germ cells). In other words: there is a far-reaching consensus *against* the removal of genetic deficiencies that is passed to the offspring.

The most important objections to those manipulations are as follows:

3.1.2.1 Objections against embryo experimentation. Before germ-line gene therapy could be *performed*, it would have to be *developed*. Its development would involve a massive experimentation with a great number of *in vitro* fertilized eggs or of embryos in the first stages of cell development, some of which would have to be produced merely for this purpose, and most of which would not survive. This cannot be justified.

This objection is voiced by politicians,[18] feminists,[19] both Protestant and Catholic theologians,[20] lawyers,[21] and philosophers.[22]

The ban on experimentation with human embryos is now firmly anchored in German law. Until recently, the law saw embryos as 'penally irrelevant objects of experimentation'. This, however, was seen as an undesirable legal loop-hole. The resulting *First Draft of a Law for the Protection of Embryos*[23] (henceforth: *DEGSE*) proposed that the fertilization of human eggs for purposes other than their transfer to a woman should be punishable by up to three years' imprisonment or a fine (*DEGSE*, p. 90). What was for a long time under dispute was whether research involving surplus embryos (among the embryos obtained in the process of *in vitro* fertilization, surplus embryos are those neither transferred to the uterus nor freezed for a future transfer) should be permitted under certain conditions. Among others, the following conditions were discussed: that the 'parents' gave their assent to experimentation; that the results could be obtained only by embryo experimentation; and that the experiments served to save other human lives.[24]

Compared to the legislation in other countries, the *Law for the Protection of Embryos*[25] (*GSE*), which became effective in January 1991, is extremely restrictive. Not only does it prohibit the sale, use or acquisition of *in vitro* fertilized eggs for all purposes other than the induction of pregnancy (*GSE*, § 2),[26] but it also prohibits the generation of more than three embryos per IVF attempt (*GSE*, § 1, sections 3–5). Surplus embryos can therefore hardly come into existence lawfully at all, and may, in any case, not be used for research.

Most discussions convey the impression that characterizing an action as an experiment with human beings is doing more than enough to disqualify it morally. See, for example, Hans Jonas:

> Experiments on unborn children are *in themselves* unethical.
>
> (Jonas 1984a: 14; my italics)

This is a little surprising. First, there has never been a moral system (in either philosophy or religion) that included 'Do not experiment upon thy fellow-men' (or, as would be needed to cover the embryo case, '. . . upon thy *potential* fellow-men') among its basic axioms.

Second, the absence of such an axiom seems quite reasonable; for there are large classes of experiments on human beings (in psychology, or in testing new medicaments) whose moral permissibility nobody doubts. What's wrong with finding out how the foetus reacts to *The Little Serenade*, or how he recognizes his mother's voice? Jonas just *cannot* mean what he says. It must be *something about* experiments on embryos that makes them wrong, and it must be something about only *some* such experiments.

3.1.2.2. The argument from human dignity. One candidate for such a 'wrong-making feature' of some experiments is their violating human dignity (*Menschenwürde*):

> Experiments [on human embryos] [. . .] are research consuming embryos. Only a *prohibition* can prevent that violation of human dignity. Where the violation of human dignity starts, it sets absolute limits to the freedom of research, if that freedom applies at all to the experiments in question.
>
> (Vitzthum 1987: 256f.)[27]

Human dignity is (like sauerkraut) not exclusively German, but a German speciality. The reason is that it figures at a prominent place in the German constitution: 'Human dignity is inviolable' is the constitution's first sentence, and therefore the term looms large in our country's legal and (as a consequence, in its) moral discussion and consciousness.

In order to find out which actions respect for human dignity does or does not permit or prescribe, we have to know what human dignity *is* or, in other words, what 'human dignity' *means*. What we find in the relevant canonical interpretations of the constitution[28] is an enumeration of types of actions that are held to be invasions of human dignity. Actions of these types are said to reduce man to a mere means.

That wording occurs frequently and is a reference to another dominant and much-quoted landmark in German moral and legal reasoning: Immanuel Kant. 'Man', says Kant in the *Groundwork of the Metaphysics of Morals (GMS)*,[29] 'and in general every rational being, exists as an end in himself, *not merely as a means* for arbitrary use by this or that will' (*GMS*, p. 428/95), and in so far he has in himself 'an absolute value' (*GMS*, p. 428/95), an 'intrinsic value – that is, dignity' (*GMS*, p. 435/102). According to Kant, human nature, thus defined, yields the categorical imperative: 'Act in such a way that you always treat humanity, whether in your own person or in the person of any other, never simply as a means, but always at the same time as an end' (*GMS*, p. 429/96).[30]

What the principle of human dignity forbids, then, is making man a mere means. Yet what does it mean to make someone a mere means? It means to ignore his ends; to treat his preferences as irrelevant.[31] But embryos do not *have* preferences; *a fortiori* none that we could ignore: hence, *whatever* we do with embryos, we do not thereby violate their human dignity, for it *cannot* be violated.

It may be objected that embryos have *potential* preferences, and that it is our duty both to see to it that these preferences become actual (which they don't if the embryo is destroyed in an experiment) and that they are satisfied. But if we had a duty to actualise potential preferences, we would also have a duty not only to let foetuses survive but to bring them into existence, a duty, that is, to procreate. Everbody denying that duty (and that means almost everybody *tout court*) must deny an embryo's right to life as well.[32]

3.1.2.3. The argument from self-determination. But what if the embryo is not destroyed? Can one not violate somebody's human dignity if in his embryonic state one does things to him that cause him to be later, in his adult life (when he *has* preferences), deeply unhappy? One certainly can. Here (when we look at the *future adult's* preferences) the principle of human dignity indeed forbids some types of manipulations – namely those rendering the future person unhappy. However, it speaks *in favour* of others – namely those rendering the future person happier, for example healthier, than he would have been if 'un-manipulated'.

Unfortunately, many people miss the last bit. Thus, philosopher Walther Chr. Zimmerli:

> today we living individuals [would] thereby present entire generations after us with an altered genetic structure . . . about which they had not been consulted. It is, therefore, as it were, a self-experiment with the forced participation of future generations, in which the fundamental right of every man to be consulted whenever he is personally affected, a fundamental right not disputed in any ethic, would be eliminated.
>
> (Zimmerli 1985: 78)[33]

Was Mr Zimmerli consulted about his own genetic make-up? Or, for that matter, about his being born at all? If not, does he think of this as an injury to his fundamental right to autonomy? It certainly violates 'the fundamental right of every man to be consulted whenever he is personally affected'. It looks as if the pre-natal observance of this right is a pretty general problem.

Given that self-determination is not at all at issue (because it is equally absent in *all* cases of bringing people into existence), why not make the best of the unavoidable fact of non-self-determination? Suppose the medical techniques were at your disposal; suppose you had refused to use them and your daughter had therefore been born with a serious, now incurable defect that you knew you could have avoided. 'Mummy', she will ask one day, 'why did you prefer my being ill to my being healthy?' 'Respect for your self-determination, my love'.[34]

3.1.2.4. The slippery-slope argument. Germ-line gene therapy is the first step towards the morally reprehensible breeding of human beings.

Hans Jonas, for example, admits that gene therapy of germ-line cells pursues *some* worthy ends (see Jonas 1984a: 14). But in performing it,

> we would open up a Pandora's box of melioristic [What's wrong with making things better?], unpredictable [see 3.1.2.5 below], inventive [What, like da Vinci's? God help us.], or simply perverse-curious adventures, abandoning the conservative spirit of genetic repair for the path of creative arrogance. We are not authorized to do this [By whom? Who authorizes me to wear socks? If nobody, is it forbidden?], and we are not equipped for it – not with the wisdom, not with the knowledge of value, not with the self-discipline. And no longer will a tradition of reverence protect us, the demystifiers of the world, from the enchantment of thoughtless crime. Therefore, let the box remain unopened.
>
> (Jonas 1984a: 14)[35]

Slippery-slope arguments adduce undesirable scenarios from the bottom of an alleged slope. But if the scenarios' badness is sufficiently obvious *for the argument to work*, it is also sufficiently obvious *for us to avoid them* and to rule them out by law. Two-thirds of the way down the slope, actions may be waiting for us that are bad too, but less obviously so. But then we can avoid the doubtful area as well and remain in the realm of the good. Hence, failure to do the obviously good cannot be justified the way Jonas tries to justify it.[36]

3.1.2.5. The argument from unpredictable consequences. According to this (less frequent) argument:

> A germ-line gene therapy must be unequivocably rejected because of the unpredictable consequences for the individuals and their offspring.
> (*Rules concerning Gene Therapy with Human Beings*, § 3)[36a]

Predictability, however, is a matter of more or less. (If I play dice, for instance, I do not know *which* number will come up, but I still know it will be a natural number between one and six.) And in many contexts (like the one at hand), *precise* prediction of the outcome is not necessarily important. What counts is whether there is a genuine risk of a *morally problematic* consequence. But this is not obviously so (if we discover the embryo to be a monster, we can still destroy it, which is in itself morally harmless: see above, section 3.1.2.2). Note also that, like so many other arguments reported here, this one is, once it has spotted a risk or a disadvantage, desperately oblivious of weighing it against possible or even certain benefits. We are talking about *medical* experiments, i.e. experiments aimed at relieving suffering. If they involve suffering at all (which is quite doubtful), then it's still only suffering against suffering, and there is no way round asking *how much* suffering could be relieved, and *how much* could be caused.

3.2 Eugenic Engineering

'Eugenic genetic engineering' or 'positive eugenics' means the 'improvement' of complex human traits, each of which is coded by a large number of genes (for example intelligence and formation of the body), by manipulating particular genes in somatic cells or in germ cells, by asexually reproducing a human organism that exhibits the 'positive' traits (cloning) or by producing crosses between humans and animals (hybrids and chimeras[36b]).[37]

By and large, the German public reject this possibility even more radically than that of germ-line gene therapy.[38] Most of the discussion on eugenics is not so much an argumentative critique as a purely rhetorical exercise, where what counts is insults rather than reasons. 'Eugenics' itself is in most contexts used as if it were an invective. Equating 'eugenics' and 'breeding' brings to mind pictures of our treating humans like cattle. Talk of the 'hybris of desiring improvements' (Broch 1989: 404), the 'path of creative arrogance' (Jonas 1984a: 15) and the 'boundless over-estimation of self-importance' (Eibach 1983: 182) replaces talk of pros and cons. Talk of a 'disguised attack on the life of handicapped people' (Aurien 1990: 55) or simply 'the "new" Eugenics' (Goettle 1990: 70) insinuates a relevant resemblance between eugenics and those Nazi crimes euphemistically called 'eugenics' by the fascist propaganda machine.[39] Together with technologies in reproduction, eugenics is considered to be not only racist, but also sexist and a 'patriarchal war against women' (Mies 1986: 44). More explicitly:

> The possibilities of a new eugenics on a world scale, a scale that would make Hitler's racial policy look like a children's game, are not the unwanted by-products of genetic and reproduction technologies; rather, these possibilities are at the core of such technologies. For if the aim were not a systematic policy of elimination and selection, to what end does one manipulate the gene at all? Life, and now man as well, is to be adapted to the necessities of the industrial system. What we know of the respective developments in the USA suffices to recognise the relationship between sexism, racism, and these technologies.
>
> (Mies 1986: 45ff.)

But we should not forget to mention the few endeavours to actually *argue* the case against eugenics.

3.2.1. Again, various forms of the argument from human dignity are to be found.

3.2.1.1. We are already acquainted with the most important type; it is by being made a mere means that a person has his dignity infringed. Since we have already discussed this type of argument in some detail (see 3.1.2.2), we may be brief here. Philosopher Günther Patzig writes:

> It would be a deviant and, clearly, a morally reprehensible idea to 'breed' . . . a class of non-aggressive, mentally very limited individuals who would probably be entirely happy with industrial working conditions intolerable for man as we know

> him. Yet such a production of human beings for the ends of other human beings would clearly go against the principle of human dignity. Here human beings would indeed be made mere means. This would also be an infringement of justice, in so far as the individuals created could not agree to such a procedure and would even have to reject it decidedly.
>
> (Patzig 1988: 36)

The first thing one notices here is the rash step from 'for the ends of others' to (my emphasis) '*mere* means'. In general, that step is as valid as inferring from the existence of apples that all there is is apples. Moreover (and more specifically), in Patzig's scenario, the new guys *are* 'entirely happy': so where have their ends been ignored? And, for the same reason, how do we know that (even in a hypothetical choice situation) they would reject that arrangement?

3.2.1.2. Thomas Broch (who at that time ran the public relations department of the largest German Catholic charity organization, the Deutscher Caritasverband), must face a similar question to the one we put to Zimmerli (see section 3.1.2.3), when he writes:

> In the area of genetic technologies, that verdict[40] also rules out . . . all measures related (one way or the other) to the improvement, that is, to the '*eugenics*', *of human life*, indeed all measures presuming to breed or construct (in accordance with any criteria whatsoever) 'more ideal' or just 'more useful' people. The most noteworthy case this applies to is the development of human–animal hybrids. But the *transfer of information into gametes* is subject to that evaluation, too . . . To a degree, cases of *in vitro* fertilization with the biological parents not knowing each other imply such a hybris of wanting to improve things. It goes without saying that the *cloning* of human individuals falls within this judgment. . . . All these are flagrant cases of control and determination of future human beings' identity. . . . The creative will of people alive today would exert power upon the personal essence and identity of future individuals, without these individuals' having a possibility to receive reparations, or an opportunity to call those who have misused their power to account.
>
> (Broch 1989: 403f.)

All this, of course, holds true for ordinary old-fashioned love-making run-of-the-mill parents as well.[41]

3.2.1.3. A different type of the argument from human dignity goes back to Hans Jonas and is an argument, not against positive or negative eugenics in general, but against cloning. It amounts to claiming a sort of right to uniqueness. As lawyer Arthur Kaufmann puts it:

> A man who is conscious of the fact that he has already existed as someone else has all naturalness, all spontaneity, all the unburdened who-knows-whence? of existence taken away from him, for his fate stands before him like a mirror before his eyes. He is robbed of the primordially human experience of freedom . . ., because, to refer to Hans Jonas, man can be free only to the degree to which he does not know his own fate (as far as it is genetically determined).
>
> (Kaufmann 1985: 272f.)[42]

If *consciousness* of non-uniqueness were the moral problem, one might be tempted to object, then why not solve it by hiding the relevant facts from the cloned offspring? This, however, would involve major technical and moral problems, most notably the frustration of people's desire to know who they are and where they come from. Secrecy or deceit are no way out, then, and Kaufmann seems to be justified in disregarding these options altogether. The question remains how Kaufmann knows that people *want* to be genetically unique. *If*, but *only if*, they do, then cloning violates somebody's right to uniqueness.

3.2.2. The doctrine of the inviolability of human nature is an endeavour to argue the case against eugenics in general. The Evangelical Church of Germany, for instance, considers cloning, as well as the construction of chimeras and hybrids, as morally prohibited, because 'the *given form of human life*' (EKD 1987: 126; my italics) is thereby injured. For lawyer Erwin Deutsch, any experiment attempting to produce a combination between an animal and a human being contradicts the *essence of humanity* and is, accordingly, 'unethical' (Deutsch 1985: 93). And philosopher and lawyer van den Daele expresses his misgivings as follows:

> If we leave man as he is, then we probably miss out the

> technical possibilities of problem-solving in a few cases. But we can't be completely wrong. And there is no moral liability for abstaining from reform of human nature.
>
> (van den Daele 1985: 209)

But the doctrine of the inviolability of human nature is problematic. Attempts to spell it out tend to yield either absurd or pointless versions. If, for example, we take the doctrine to forbid any interference with human nature, this will rule out not only medicine but most forms of environmental and social changes. If, to quote another example, we take it just to forbid the direct alteration of genes, this rules out negative as well as positive eugenics. (Even if you welcome that result, the question remains where you get that prohibition *from*. You hardly want 'Don't fiddle around with genes' to be a moral *axiom*, do you? If such fiddling is wrong, then, presumably, it is because there is *something about* it that *makes* it wrong. But what?) Perhaps, in the end, we don't want the doctrine to mean anything but a ban on positive eugenics. But if it's synonymous with the ban, it can hardly justify it.

To the contrary, it is then in need of a justification as much as the ban itself. It is far from obvious that we must preserve all the characteristics that are natural to us, such as vulnerability to sickness, aggression, brutality and insincerity. Why should it be of intrinsic worth to protect these attributes? Imagine that by snapping your fingers, you could bring about a world in which human nature was changed, a world without these attributes. What would be wrong with that?

This is not to say that there are no good arguments against positive eugenics at all, but only that the inviolability of human nature itself is in need of an argument. Perhaps the most plausible argument against positive eugenics rests on a general objection to any group of people trying to plan too closely what human life should be like.

> The present genetic lottery throws up a vast range of characteristics, good and bad, in all sorts of combinations. The group of people controlling a positive engineering policy would inevitably have limited horizons, and we are right to worry that the limitations of their outlook might become the boundaries of human variety.
>
> (Glover 1984: 47)

And even if a genetic supermarket could solve *this* problem, it would seem to be highly probable that human variety would be lost, for instance by fops. These are genuine dangers, constituting one genuine counter-argument. But the inviolability of human nature plays no part in it, nor should it:

> Preserving the human race as it is will seem an acceptable option to all those who can watch the news on television and feel satisfied with the world. It will appeal to those who can talk to their children about the history of the twentieth century without wishing they could leave some things out.
>
> (Glover 1984: 56)

4 CONCLUSION

Of course, there *are* proponents of research into (and, in case it works, application of) most forms of genetic technologies in Germany as well. They are in a minority. They see weaknesses (some of which I have tried to point out above) in their opponents' arguments; given a specific procedure, experiment or technology, they are ready to weigh its risks against its possible benefits, and in some cases reach the conclusion that, all things considered, we ought to try it. Both their critical and their constructive arguments are extremely similar to those of the leading English-language philosophers working in the field (see, for instance, Glover 1984; Harris 1992; Singer and Wells 1984). We touched only briefly on *these* arguments because they were outside the scope of this chapter; we thus had to look primarily at the main manoeuvres *against* most forms of genetic engineering, and at their sources. Hardly anywhere else is the public's opinion against the technologies in question as united and as strong as in Germany, and hardly anywhere else is the pertinent legislation as restrictive. The forces we find are these: the influence of the German constitution (hence the ubiquitous references to 'human dignity'); the influence of Kant (hence 'autonomy', 'end in himself', 'mere means', 'categorical imperative'); the frightful tendency to firmly adhere to simple, rigorous rules of thumb ('whatever the consequences', which is again partly due to Kant); the influence of centuries of hazy pseudo-rational philosophy (hence: more poetry than reasoning, more rhetoric than logic, and more quotations than arguments); the Christian churches' unbroken influence on morals and public affairs, including legislation (hence: premises, moral judgments and legal regulations, for example on the unconditional protection of all forms of human life, which both stem from religion and can hardly be justified without it); and an unreasonable way of adducing Germany's Nazi past. (No doubt maximum care should be taken to remember Nazi crimes and to avoid their repetition; maximum care also to remember its pseudo-scientific and pseudo-ethical verbal roots and upshots. But from 'Hitler said that *p*' one cannot conclude that *p* is false. What if *p* were '2 + 2 = 4'?) – so much for the main manoeuvres and their sources.

NOTES

I am grateful to everyone who helped to make this chapter better than it would otherwise have been, particularly to Christoph Fehige, for making several suggestions and comments and going through various drafts, and to Andrew Wilson, for translating it from German into English.

1 Bundesministerium für Forschung und Technologie.
2 *Sicherheitsrichtlinien für Forschungsarbeiten über die In-vitro-Neukombination von Nukleinsäuren.*
3 *Richtlinien zum Schutz vor Gefahren durch die in-vitro-neukombinierten Nukleinsäuren.*
4 Deutsche Forschungsgemeinschaft.
5 *Chancen und Risiken der Gentechnologie*. The committee consisted of nine German Members of Parliament and eight experts; there was no philosopher. Its objective was described as the following: 'to discover the main problem areas in contemporary genetic technologies and related bio-technical research. . . . Whereby economic, ecological, legal, and social consequences should stand in the foreground. In addition the ethical aspects of the shadowy area of genetic technologies' application to humans should be taken into special consideration' (*Bericht der Enquete-Kommission 'Chancen und Risiken der Gentechnologie ' des 10. Deutschen Bundestages*, p. 1).
6 *Bericht der Enquete-Kommission 'Chancen und Risiken der Gentechnologie' des 10. Deutschen Bundestages.*
7 *Gesetz zur Regelung von Fragen der Gentechnik.*
8 Bundesministerium für Jugend, Familie, Frauen und Gesundheit.
9 Zentrale Kommission für Biologische Sicherheit.
10 The committee is composed of ten experts as well as one union representative, one member of the industrial safety organizations, and one representative each from the fields of business, environmental protection and funding. The members serve for three years. They are appointed by five ministries, under the superintendence of the BMJFFG. The members act independently and are not bound by directives of any kind.

11 The requirements increase according to the degree of danger involved in research and production. In certain circumstances, every step of the research must be finely recorded so that, even years later, the entire research and production process can be fully reconstructed.

12 Security level 1 comprises work in genetic technologies that, according to the standard of science (sos), is of no risk to human health or the environment. Security level 2 includes work that, according to the sos, involves a mild risk for human health or the environment. Security level 3 consists of work that, according to the sos, is moderately risky for human health or the environment. Activities that are (or can, on good grounds, be suspected to be), according to the sos, highly risky to human health or the environment, are on level 4.

13 Members of the public are allowed to ask questions at official hearings if a laboratory or a production plant concerned with genetic technologies is to be built in their area. See *GenTG*, § 18.

14 In 1989, the German Association of Physicians (Bundesärztekammer) put things similarly in their policy statement *Rules concerning Gene Therapy with Human Beings* (*Richtlinien zur Gentherapie beim Menschen*) (*RGTM*): 'Somatic gene therapy represents a special form of substitution therapy. . . . Although this therapy commences on the level of inherited information, intrinsically it poses no new ethical problems because it is limited in its effects to the patient who is being treated' (*RGTM*, § 2.2).

Assuming that the research work towards such gene therapy leads to a good therapy-scheme with high success rates, many people feel that somatic gene therapy is actually less problematic, from a certain point of view, than organ transplants. In so far as the implantation of genetically altered somatic cells in patients from whom they were initially taken is a particular form of auto-transplant, the rejection syndrome characteristic of heterologous organ transplants cannot occur; nor is there a donor problem. See *BCRG*, p. 183.

15 See Hülsmann and Koch 1990: 35, as well as Vitzthum 1987:272. See also the *Rules concerning Gene Therapy with Human Beings (Richtlinien zur Gentherapie beim Menschen)*, § 2.3.

16 See also Rainer Hohlfeld (from the Hamburg Institute for Social Research (Hamburger Institut für Sozialforschung): genetic engineering is the 'ultimate logical conclusion of scientific reasoning and experimental biomedical research' (Hohlfeld 1989: 170), which reduces the cause of the illness to one single factor, namely the 'biomedical effect'. 'If biomedical research proceeds that way, its construction of reality is, therefore, a-ecological, ahistorical and asocial. The phenomena of life have a dimension that is psycho-social, ecological, subjective and linked to natural history; modern medicine refuses to see this dimension and for its resulting blindness is correctly criticized today as insufficient' (Hohlfeld 1989: 179).

See also the resolution passed by the participants of the 'Women against Genetic and Reproduction Technologies Congress': 'Genetic and reproduction technologies offer medicine a growing arsenal of methods by which to fight symptoms. The principle of "self-imposed" illnesses is placed at the forefront. Hereditary factors are made responsible for people responding to progressive environmental destruction and pollution with illnesses (*Resolution*, p. 17).

17 See Bayertz 1991: 302 and van den Daele 1985: 187.

18 See, for example, the majority vote of the Enquete Commission: 'The effective development of germ-line gene therapy currently presupposes an experimental "consumption" of embryos that can under no circumstances be accepted,' (*BCRG*, p. 189). The passage is accompanied by a reference to a part of the World Medical Assembly's declaration (Helsinki/Tokyo, 1975), saying that in experiments on human beings concern for the interest of the 'test subject' must always prevail over the interests of science and society.

19 See, for example, Mies 1986: 98, on genetic technologies and embryo experimentation: 'What kind of research is it that wants to slaughter . . . human beings for the sake of science and its own ends? I call it a *cannibalistic research*, although I am aware of the fact that I probably do the cannibals an injustice. . . . What is clear, at any rate, is that they [the scientists] will have to cast all scruples concerning the integrity of persons aside, if they want to have sufficient "raw material" for their extravagant research.'

20 The Evangelical Church of Germany (Evangelische Kirche Deutschlands (EKD)) explained: 'Deliberate manipulation of human embryos that would anticipate the destruction of such embryos is not morally tenable,' (*Announcement of the 7th Synod of the Evangelical Church of Germany* (*Kundgebung der 7. Synode der EKD*), p. 126).

At the national level, the Catholic Church has not made any similar statement. But in the *Instruction concerning the Respect for the Beginnings of Human Life and the Dignity of Reproduction* (*Instruktion über die Achtung vor dem beginnenden menschlichen Leben und die Würde der Fortpflanzung*) of 3 October 1987, the Vatican Congregation for the Propagation of the Faith declared: 'No projected goal, no matter how noble such a goal, as for example its future value for science, other human beings or society, can justify any form of experimentation with living embryos or foetuses, whether they are capable of life or not, whether they are in vivo or in vitro,' (*IALWF*, p. 17).

21 See the report of the civil law section of the 56th Lawyers' Society Congress (*Juristentag*), the regular meeting of the German lawyers: 'There is unanimity . . . that it should only be legal to generate embryos if this is not done for research purposes and if their subsequent reimplantation and their full development to human life are aimed at' (Franzki 1987: 39).

22 See, for example, Reinhard Löw: 'No matter how worthy the goal, it cannot heal a means that is inherently bad – as the intervention in the personality of a human life' (Löw 1983: 44). It is clear from the context that by 'intervention in the personality of a human life' he refers to embryo experimentation. See also Birnbacher 1987: 83; Wimmer 1990: 66.

23 *Diskussionsentwurf eines Gesetzes zum Schutz von Embryonen*. That draft was based on the report of what became known as the Benda

Commission (Benda Kommission). In May 1984, a working party, '*In vitro* Fertilization, Genome Analysis and Gene Therapy' (*In-vitro-Fertilisation, Genomanalyse und Gentherapie*) was set up by the Federal Ministery of Justice (Bundesjustizministerium) and the BMFT, under the direction of Ernst Benda, former president of the Federal Constitutional Court (Bundesverfassungsgericht). The Benda Commission concerned itself in particular with the legal and ethical questions stemming from any of the above-noted techniques and, in its final report, made large numbers of suggestions for possible legal measures. See *In Vitro Fertilisation, Genome Analysis and Gene Therapy. Report of the Working Party of the Federal Minister for Research and Technology and of the Federal Minister of Justice (In-vitro-Fertilisation, Genomanalyse und Gentherapie. Bericht der gemeinsamen Arbeitsgruppe des Bundesministers für Forschung und Technologie und des Bundesministers der Justiz).*

24 Such conditions were proposed, for instance, by the Benda Commission (see note 23 above), §§ 2–3 of its report (see note 23); see also the *Regulations for Embryo Experimentation* (*Richtlinien zur Forschung an frühen menschlichen Embryonen*) of the German Association of Physicians (Bundesärztekammer), §§ 3.1.1–3.1.3; the *Report of the Interministerial Commission for the Clarification of Bioethical Questions* (*Bericht der interministeriellen Kommission zur Aufarbeitung von Fragen der Bioethik*), thesis IV. (The interministerial commission was set up in 1985 by the Rhineland Palatinate Minister of Justice.)

The Max Planck Society (Max Planck-Gesellschaft) (MPG), asked to comment on the *DEGSE*, pleaded for the lawfulness of embryo experimentation (as a part of academic freedom). See Hofschneider 1989: 14. One year later, however, the MPG revised its position, reacting to the angry critique by the public. It decided to do without embryo experimentation. See Meermann 1988: 9–11 and Hofschneider 1989: 16.

For restrictive regulations (or pleas for them), see the *General Policy of State Regulations concerning Reproductive Medicine* (*Gesamtkonzept staatlicher Maß nahmen in der Fortpflanzungsmedizin*); the *Proposals on Reproductive Medicine and Human Genetics from the German Judges' Organization* (*Thesen des Deutschen Richterbundes zur Fortpflanzungsmedizin und zur Humangenetik);* § 2 of the *Second Draft of a Law for the Protection of Embryos* (*Arbeitsentwurfs eines Gesetzes zum Schutz von Embryonen*), of the *Third Draft of a Law for the Protection of Embryos* (*Entwurf eines Gesetzes zum Schutz von Embryonen*), and of the *Law for the Protection of Embryos* (*Gesetz zum Schutz von Embryonen*).

25 *Gesetz zum Schutz von Embryonen.*

26 More explicitly: '(1) Whoever sells, transfers or acquires an *in vitro* fertilized embryo or an embryo that was extracted prior to its attachment to the uterus, for purposes other than its preservation, will serve up to three years in prison or will pay a monetary fine. (2) Whoever effects the extra-uteral development of a human embryo for purposes other than the induction of a pregnancy will be subject to punishment. (3) Intent is also punishable,' (*GSE*, § 2).

27 See also philosopher Reiner Wimmer, who says that the categorical imperative asks us to respect the autonomy and dignity of man; and that, accordingly, it forbids the use of man as a mere means. 'In my opinion total instrumentalization occurs if parents, experimenters, or others dispose of a human life at will – even when it is an early life, one without personhood but with the potential to it' (Wimmer 1990: 63).

Similarly, the Vatican: 'To use the human embryo or the foetus as an object or a means for experiments is a crime against their dignity as human beings, for they are due the same right and respect as a born child and every human person' (*Instructions concerning the Respect for the Beginning of Human Life and the Dignity of Reproduction* (*Instruktion über die Achtung vor dem beginnenden menschlichen Leben und die Würde der Fortpflanzung*), p. 17).

And lawyer Moni Lanz-Zumstein declares: 'It would contradict the constitutional guarantee of human dignity, and the spirit of this fundamental value, if embryos were produced for mere research purposes. . . . The objectivation and determination of human life as a mere means to an end is manifested here in its most extreme form. Human life would serve here as a thing freely disposed of and extrinsically determined, to be used for the purposes and goals of others. Not even the highest research goals can justify the artificial creation of human beings, regardless of whether the experiment is consuming embryos or whether it is attempting to keep them alive artificially' (Lanz-Zumstein 1986a: 105).

28 Cf., for example, Maunz/Dürig 1989, article 1, § 1, 28.

29 *Grundlegung der Metaphysik der Sitten*.

30 More explicitly: in *GMS* Kant tries to prove *a priori* that there actually is an absolute practical law, the categorical imperative. He says that *if* there were something which could be the ground of a possible categorical imperative, it would be something whose existence has in itself an absolute value, something which has an end in itself. (Ends that a rational being adopts as effects of his actions (material ends) are only relative: it is merely their relation to the subject's preferences that gives them their value.) And then Kant says: in fact, there is something which has an end in itself: it is man, and in general every rational being. *Ergo*: the categorcial imperative. Of course, for the inference to be valid Kant should have said: everything that has an end in itself is the ground of the categorical imperative.

31 See Hare 1993. For a weaker explanation going roughly in the same direction as ours, see philosopher Norbert Hoerster: 'Human dignity is *not* . . . a given, recognizable something (as, for example, with human life) that allows an objective determination of which actions harm or protect it. To be sure, the concept of human dignity is not of a *purely* normative nature; . . . it has . . . a descriptive element, namely that man is by nature capable of self-determination. The unavoidable and decisive questions that determine the meaning of the term "human dignity", i.e. those asking which forms of self-determination are *morally legitimate* (whether, for example, murder, the death penalty, suicide, bodily injury, sale of labour power, sale of sexual services, polygamy, sodomy,

abuse of animals [are morally legitimate]), are and remain questions of *value*' (Hoerster 1983: 96).

32 For a more detailed discussion on merely potential preferences see Wessels 1994, and *Verbietet das Recht auf Leben Abtreibung?* by Wessels.

33 From this, Zimmerli 'concludes': 'Specific gene transfer into germ cells must be *banned* . . . – even if this means that illnesses cannot or can only partly be avoided prophylactically' (Zimmerli 1985: 79).

See also Reinhard Löw, quoting Robert Spaemann: 'Our technical "know-how" (here our know-how in genetic technologies) . . . leads to a constantly growing power over the coming generations, and thus, from their point of view, to a domination of the dead over the living. This is an example of the "right of might", i.e. of injustice' (Löw 1983: 42). See also Birnbacher 1987: 82f.

Similar to the above is Benda 1985: 227: 'A gene transfer into germ cells means . . . a determination of the progency by others, i.e. by parents, scientists or the state.'

34 See also philosopher Hans-Martin Sass: 'Therapy dealing with the severest of inherited mental diseases already recognizable in germ cells is not only morally acceptable but morally required. The withholding of such therapy would be morally reprehensible; it goes against human dignity, the obligation to be responsible to one's neighbour, and runs contrary to one's own conscience' (Sass 1987: 92).

35 Note that, according to Jonas, the thesis that breeding human beings should not be allowed is based on our 'duty to the existence and *essence* of future generations' (Jonas 1984b: 86; my italics). (There is even more poetry to come when Jonas 'argues' against alterations whose results may not be human beings any more. His rejection is then based upon the 'idea of man as one which demands its manifestation in the world', and upon the 'categorical imperative that there be people at all' (Jonas 1984b: 86ff).)

In comparison to Hans Jonas, see the less emotive statement of the Enquete Commission: 'Every introduction or use of gene manipulation on germ cells would certainly begin with the treatment of illnesses whose evaluation is widely agreed upon in society. But it would not necessarily remain restricted to this. If the technique of genetic correction in the germ line is established, the transition to improvement and breeding will become fluid. There is already a "grey area" in the concept of illnesses. With such attributes as small body size, low intelligence quotient, inclination to depression or displays of rage, and so forth, it is unclear when such attributes are peculiar to the individual within the broad range of natural diversity and when they are pathological. Should it be found that such attributes are, entirely or partly, genetically determined and can be influenced, the border between medically legitimate correction and breeding would be easy to shift' (*BCRG*, p. 189). See also lawyer and philosopher van der Daele: he emphasises, as does philosopher Kurt Bayertz (1990: 4), that germ-line gene therapy cannot, *by itself*, be evaluated differently from somatic gene therapy. But 'gene therapy of germline cells is nearer to the danger of human breeding than is somatic cell therapy' (van den Daele 1985: 197). Second, avoidance of misuse via the complete abandonment of the technologies would be safer than control of its application. In view of this, 'it would be better to go completely without gene manipulation, if the medical options that are thereby deleted are not extraordinary and irreplaceable. But exactly this they don't seem to be' (van den Daele 1985: 197).

36 See also Dieter Birnbacher's objection to this kind of slippery slope argument: it is quite doubtful whether the risks involved in the development of germ-line gene therapy can outweigh the benefits it promises (Birnbacher 1989: 218).

36a *Richtlinien zur Gentherapie beim Menschen*.

36b In general, a hybrid is an organism (or a piece of DNA) constructed from the genetic material of two different species. The word 'chimera' is often used synonymously. But, in fact, it refers to hybrids which are clearly identifiable as odd, as the sort of creature that one would call a monster.

37 In contrast, negative eugenics is the elimination of particular 'negative' traits, for instance by sterilizing whoever exhibits, or aborting whoever *would* exhibit, such traits.

Conceptually, the distinction between positive and negative eugenics is a little suspect. If I *reinforce* property *F*, I can always say I *eliminate* property *LACKING F*. (If in the previous sentence '*reinforce*' and '*eliminate*' change places, we get the same trick the other way round.) Notice that this has nothing to do with a slippery-slope problem: the trick works even in the clearest cases. The moral is that we should not take the expressions 'reinforce "positive" traits' or 'eliminate "negative" traits' at their descriptive face value, but should read them as referring each to a list of specific actions (a list 'making people more beautiful, more intelligent (etc.)' and a list 'rendering people immune to cancer (etc.)') where the two lists do not overlap. And this is indeed what everybody does before he notices the trick.

38 This tendency has already been visible (section 3.1.2). The fluid transition between healing and breeding was reported there to be considered as a major argument against gene therapy.

39 Sometimes, however, it does not stop at this allusion, at least not with Hans Jonas: 'After its terrifying experiment in recent German history, we do not need to deal in any detail with positive eugenics as a systematic human selection with the goal of the improvement of the species. Its moral and political offensiveness need no exposition in this country' (Jonas 1987: 176). See also Aurien 1990: 49: prenatal diagnosis assists 'an old eugenics in new clothes to take the stage once more'. See also Eibach 1983: 174ff.

40 The following verdict is meant: 'Man should never be allowed to become an instrument, a means for the ends and interests of others. In other words, man within the undecipherable, inscrutable entirety of his personal identity and uniqueness, within his philanthrophy and sociality, should not be allowed to become functionalized and instrumentalized' (Broch 1989: 403).

Note that here the 'mere' that even Kant remembered is omitted right from the start. Thus, Broch's verdict forbids you to give anybody a helping hand.

41 See also Hans-Martin Sass's impressive 'inference' by means of a non sequitur, that concerns – among other things – the same critique: '*If* we accept, even promote, on the one side, that there should not be any indoctrination of human beings, *then* we would also reject positive eugenics of human beings' (Sass 1987: 104f.; my italics).

42 See also Hans Jonas himself: 'The whole thing is frivolous with respect to the motives and morally reprehensible with respect to the effects: as in the case of other biological daring deeds, here just one attempt would be frivolous.' Thus, 'knowing oneself to be an imitation of a being that already revealed itself through a life would strangle the authenticity of being oneself, the freedom to discover oneself, as well as the freedom to surprise oneself and others at what is within oneself. . . . A fundamental right of ignorance, which belongs indispensably to existential freedom, is injured here anticipatorily' (Jonas 1984a: 13).

See also Ernst Benda: 'From the point of view of human dignity, there exists an elemental claim of the growing human being not to be a copy of his parents, but his or her own unique personality. This claim is justified immediately from the essence of man. The same holds from the point of view of a people or even of humanity in its entirety' (Benda 1985: 224). See also Löw 1983: 43; Honecker 1985: 154; Broch 1989: 404.

BIBLIOGRAPHY

Reports of State Commissions

Bericht der Enquete-Kommission 'Chancen und Risiken der Gentechnologie' des 10. Deutschen Bundestages (Report of the 10th German Parliament's Enquete Commission. 'Chances and Risks of Genetic Technology') (1987), ed. by the *Deutscher Bundestag*, Bonn.

In-vitro-Fertilisation, Genomanalyse und Gentherapie. Bericht der gemeinsamen Arbeitsgruppe des Bundesministers für Forschung und Technologie und des Bundesministers der Justiz (In Vitro Fertilisation, Genome Analysis and Gene Therapy. Report of the Working Party of the Federal Minister for Research and Technology and of the Federal Minister of Justice) (1985), ed. by the *BMFT*, München.

Gesamtkonzept staatlicher Maßnahmen in der Fortpflanzungsmedizin (General Policy of State Regulations concerning Reproductive Medicine) (1986), by the *Justizministerium* of Baden-Württemberg, in and *cit. sec.* Seesing 1987: 97–8.

Bericht der interministeriellen Kommission zur Aufarbeitung von Fragen der Bioethik – Fortpflanzungsmedizin – und vorläufiger Arbeitsentwurf eines Landesgesetzes über Fortpflanzungsmedizin (Report of the Interministerial Commission for the Clarification of Bioethical Questions) (1986), by the *Ministerium der Justiz* of Rheinland-Pfalz, in and *cit. sec.* Seesing 1987: 119–41.

Laws and Drafts of Laws

Gesetz zur Regelung von Fragen der Gentechnik (*Law to Regulate Questions of Genetic Technology*) (1990), by the *Bundesjustizminister* in and *cit. sec.* Presse- und Informationsamt der Bundesregierung 1990a: 21–40.

Diskussionsentwurf eines Gesetzes zum Schutz von Embryonen (*First Draft of a Law for the Protection of Embryos*) (1986), by the *Bundesjustizminister*, in and *cit. sec.* Hülsmann and Koch 1990: 90–2.

Arbeitsentwurf eines Gesetzes zum Schutz von Embryonen (*Second Draft of a Law for the Protection of Embryos*) (1988), by the *Bundesjustizminister*, in and *cit. sec.* Hülsmann and Koch 1990: 92–6.

Entwurf eines Gesetzes zum Schutz von Embryonen (*Third Draft of a Law for the Protection of Embryos*) (1989), by the *Bundesjustizminister* in and *cit. sec.* Hülsmann and Koch 1990: 96–9.

Gesetz zum Schutz von Embryonen (*Law for the Protection of Embryos*) (1991), by the *Bundesjustizminister*, in and *cit. sec.* Presse- und Informationsamt der Bundesregierung 1990b: I–V.

Guidelines and Regulations

Richtlinien zur Gentherapie beim Menschen (*Rules concerning Gene Therapy with Human Beings*) (1989), by the *Bundesärztekammer*, in and *cit. sec.* Heerklotz 1989: 67–71.

Thesen des Deutschen Richterbundes zur Fortpflanzungsmedizin und zur Humangenetik (*Proposals on Reproductive Medicine and Human Genetics from the German Judges' Organization*) (1986), by the *Deutscher Richterbund*, in and *cit. sec.* Lanz-Zumstein 1986b: 212–15.

Richtlinien zur Forschung an frühen menschlichen Embryonen (*Regulations for Embryo Experimentation*) (1985), by the *Bundesärztekammer*, *Deutsches Artzeblatt* 82/50 (1985), 3757–64.

Resolutions

Resolution (*verabschiedet von den Teilnehmern des Kongresses 'Frauen gegen Gentechnik und Reproduktionstechnik*) (*Resolution* (passed by the participants of the Women against Genetic and Reproductive Technologies Congress)), 1985; in Die Grünen *et al.* 1986: 15–20.

The Churches' Comments

Kundgebung der 7. Synode der Evangischen Kirche Deutschlands (*Announcement of the 7th Synod of the Evangelical Church of Germany*) (1987), by the *Evangelische Kirche Deutschlands*, in and *cit. sec.* Schroeder-Kurth *et al.* 1988: 119–30.

Instruktion über die Achtung vor dem beginnenden menschlichen Leben und die Würde der Fortpflanzung (*Instruction concerning the Respect for the Beginning of Human Life and the Dignity of Reproduction*) (1987), by the *Heiliger Stuhl*, Bonn.

Books and articles

Aurien, Ursula (1990) 'Humangenetik und Ethik', in Bruns *et al.* 1990: 49–57.

Bayertz, Kurt (1987) *Genethik, Probleme der Technisierung menschlicher Fortpflanzung*, Reinbek near Hamburg.

— (1990) *Gentherapie am Menschen, Tendenzen der aktuellen Diskussion*, Reinbek near Hamburg.

— (1991) 'Drei Typen ethischer Argumentation', in Sass 1991: 291–316.

Benda, Ernst (1985) 'Erprobung der Menschenwürde am Beispiel der Humangenetik', in Flöhl 1985: 205–31.

Birnbacher, Dieter (1987) 'Gefährdet die moderne Reproduktionsmedizin die menschliche Würde?', in Braun *et al.* 1987: 77–88.

— (1989) 'Genomanalyse und Gentherapie', in Sass 1989: 212–31.

Braun, Volkmar, *et al.* (eds) (1987) *Ethische und rechtliche Fragen der Gentechnologie und Reproduktionsmedizin*, München.

Broch, Thomas (1989) 'Gentechnologie und Menschenwürde', *Caritas – Zeitschrift für Caritasarbeit und Caritaswissenschaft* 90(9): 402–9.

Bruns, Theo, *et al.* (1990) *Tödliche Ethik. Beiträge gegen Eugenik und 'Euthanasie'*, Hamburg.

Daele, Wolfgang van den (1985) *Mensch nach Maß? Ethische Probleme der Genmanipulation und Gentherapie*, München.

Deutsch, Erwin (1985) 'Artifizielle Wege menschlicher Reproduktion: Rechtsgrundsätze', in and *cit. sec.* Flöhl 1985: 232–47.

Die Grünen *et al.* (eds) (1986) *Frauen gegen Gentechnik und Reproduktionstechnik, Dokumentation zum Kongreß vom 19–21. 4. 1985 in Bonn*, Köln.

Eibach, Ulrich (1983) *Experimentierfeld: Werdendes Leben. Eine ethische Orientierung*, Göttingen.

Eser, Albin (1984) 'Genetik, Gen-Ethik, Gen-Recht? Rechtspolitische Überlegungen zum Umgang mit menschlichem Erbgut', in and *cit. sec.* Flöhl 1985: 248–58.

— (1987) 'Strafrechtliche Schutzaspekte im Bereich der Humangenetik', in Braun *et al.* 1987: 120–49.

Eser, Albin, *et al.* (eds) (1990) *Regelungen der Fortpflanzungsmedizin und Humangenetik*, vol. 1, Frankfurt.

Fehige, Christoph and Meggle, Georg (eds) (1993) *Zum Moralischen Denken*, Frankfurt.

Fehige, Christoph and Wessels, Ulla (eds) (1994) *Preferences*, Berlin.

Flöhl, Rainer (ed.) (1985) *Genforschung – Fluch oder Segen? Interdiziplinäre Stellungnahmen*, München.

Franzki, Harald (1987) 'Die künstliche Befruchtung beim Menschen – Zulässigkeit und zivilrechtliche Folgen. Bericht über die zivilrechtliche Abteilung des 56. Deutschen Juristentages', *Mitteilungen des deutschen*

Richterbundes, no. 1.

Glover, Jonathan (1984) *What Sort of People Should There Be?*, Harmondsworth.

Goettle, Gabriele (1990) 'A und B, Bein und Zeh', in Bruns *et al.* 1990: 69–78.

Hansen, Friedrich and Kollek, Regine (1985) *Gentechnologie – Die neue soziale Waffe*, Hamburg.

Hare, Richard M. (1993) 'Könnte Kant ein Utilitarist gewesen sein?', in Fehige and Meggle, 1993.

Harris, John (1992) *Wonderwoman and Superman: The Ethics of Human Biotechnology*, Oxford.

Heerklotz, Brigitte (ed.) (1989) *Biomedizinische Ethik. Europäische Richtlinien und Empfehlungen*, Bochum.

Herbig, Jost (1978) *Die Gen-Ingenieure*, München.

Hoerster, Norbert (1983) 'Zur Bedeutung des Prinzips der Menschenwürde', *Juristische Schulung* 2: 93–6.

Hofschneider, Hans Peter (1989) 'Embryonenforschung und Gentechnologie – was sagen die Wissenschaftler der Max-Planck-Gesellschaft dazu?', in *Max-Planck-Gesellschaft* 1989: 13–22.

Hohlfeld, Rainer (1989) 'Die zweite Schöpfung des Menschen – eine Kritik der Idee der biochemischen und genetischen Verbesserung des Menschen', in Schuller and Heim 1989: 228–48.

Honecker, Martin (1985) 'Verantwortung am Lebensbeginn', in Flöhl 1985: 144–60.

Hülsmann, Christoph and Koch, Hans-Georg (1990) 'Bundesrepublik Deutschland', in Eser 1990: 29–156.

Jonas, Hans (1982) 'Laßt uns einen Menschen klonieren. Von der Eugenik zur Gentechnologie', in Jonas 1987: 162–203.

—— (1984a) 'Technik, Ethik und Biogenetische Kunst, Betrachtungen zur neuen Schöpferrolle des Menschen', in and *cit. sec.* Flöhl 1985: 1–15.

—— (1984b) *Das Prinzip Verantwortung, Versuch einer Ethik für die technologische Zivilisation*, Frankfurt.

—— (1987) *Technik, Medizin und Ethik*, Frankfurt.

Kant, Immanuel (1797) *Grundlegung der Metaphysik der Sitten*, in *Werke*, vol. IV (ed. by the *Königlich Preußischen Akademie der Wissenschaften*), Berlin, 1911; (*Groundwork of the Metaphysics of Morals*, translated and analysed by H. J. Paton, New York, 1956.)

Kaufmann, Arthur (1985) 'Der entfesselte Prometheus. Fragen der Humangenetik und der Fortpflanzungstechnologien aus rechtlicher Sicht', in Flöhl 1985: 259–77.

Kluxen, Wolfgang (1985) 'Manipulierte Menschwerdung', in Flöhl 1985: 16–29.

Koslowski, Peter, *et al.* (eds) (1983) *Die Verführung durch das Machbare. Ethische Konflikte in der modernen Medizin und Biologie*, Stuttgart.

Lanz-Zumstein, Moni (1986a) 'Embryonenschutz. Juristische und Rechtspolitische Überlegungen', in Lanz-Zumstein 1986b: 93–114.

—— (ed.) (1986b) *Embryonenschutz und Befruchtungstechnik. Seminar bericht und Stellungnahmen aus der Arbeitsgruppe 'Gentechnologie' des Deutschen Juristenbundes*, München.

Löw, Reinhard (1983) 'Gen und Ethik. Philosophische Überlegungen zum Umgang mit menschlichem Erbgut', in Koslowski *et al.* 1983: 33–48.

Maunz, Th., Dürig, G., *et al.* (1989) *Grundgesetz. Kommentar*, vol. 1, München.

Max-Planck-Gesellschaft (ed.) (1989) *Respekt vor dem werdenden Leben. Ein Presseseminar der Max-Planck-Gesellschaft zum Thema Embryonenforschung* (*Berichte und Mitteilungen*, no. 4), München.

Meermann, Horst (1988) 'MPG verzichtet auf Embryonenforschung', in *MPG-Spiegel* 5/88.

Mies, Maria (1986) 'Reproduktionstechnik als sexistische und rassistische Bevölkerungspolitik', in Die Grünen 1986: 44–8.

Patzig, Günther (1988) 'Moralische Probleme der Genomanalyse/Gentherapie und In-vitro-Fertilisation', in Schlegel 1988: 30–45.

Presse- und Informationsamt der Bundesregierung (ed.) (1990a) *Schutz von Mensch und Umwelt. Das Gentechnikgesetz*, Bonn.

—— (ed.) (1990b) *Das Embryonenschutzgesetz*, Bonn.

Reiter, Johannes (1988) 'Menschenwürde und Gentechnologie', in Seesing 1988: 16–34.

Sass, Hans-Martin (1987) 'Methoden ethischer Güterabwägung in der Biotechnologie', in Braun *et al.* 1987: 89–110.

—— (ed.) (1989) *Medizin und Ethik*, Stuttgart.

—— (ed.) (1991) *Genomanalyse und Gentherapie. Ethische Herausforderung in der Humanmedizin*, Berlin.

Schlegel, Hans Günter (ed.) (1988) *Gentechnologie und In-vitro-Fertilisation. Kolloquium der Akademie der Wissenschaften zu Göttingen*, Göttingen.

Schroeder-Kurth, Traute, *et al.* (eds) (1988) *Das Leben achten. Maßstäbe für Gentechnik und Fortpflanzungsmedizin*, Gütersloh.

Schuller, Alexander and Heim, Nikolaus (eds) (1989) *Der codierte Leib. Zur Zukunft der genetischen Vergangeheit*, Zürich/München.

Seesing, Heinz (ed.) (1987) *Technologischer Fortschritt und menschliches Leben. Die Menschenwürde als Maßstab der Rechtpolitik*, part 1, München.

—— (ed.) (1988) *Technologischer Fortschritt und menschliches Leben. Die Menschenwürde als Maßstab der Rechtspolitik*, part 2, München.

Singer, Peter and Wells, Deane (1984) *The Reproduction Revolution. New Ways of Making Babies*, Oxford.

Suzuki, David T., *et al.* (1989) *An Introduction to Genetic Analysis*, 4th edition, New York.

Vitzthum, Wolfgang Graf (1987) 'Das Verfassungsrecht vor der Herausforderung von Gentechnologie und Reproduktionsmedizin', in Braun *et al.* 1987: 263–96.

Wessels, Ulla (1994) 'Midwives and rabbits. Some questions', in Fehige and Wessels, 1994.

—— (forthcoming) *Verbiteit das Recht auf Leben Abtreibung?* (unpubl. diss.).

Wimmer, Reiner (1990) 'Zur ethischen Problematik der Keimbahn-Gentherapie am Menschen', *Zeitschrift für philosophische Forschung*

44(1): 55–67.

Zimmerli, Walther Ch. (1985) 'Dürfen wir, was wir können? Zum Verhältnis von Recht und Moral in der Gentechnologie', in Flöhl 1985: 59–85.

14

GENETIC ENGINEERING IN THEOLOGY AND THEOLOGICAL ETHICS

Anthony Dyson

INTRODUCTION

My purpose in this essay is to give some account of the ways in which Christian theology and Christian ethics try to contribute to the debate about genetic engineering. There is in fact a large corpus of material on this topic, but some of it is repetitive. For the most part I shall analyse and quote from two authors and from a small number of 'church reports' which effectively point both to divisions of opinion and to consensus about resources and methodology. I am well aware that more, or different, reports could have been called in evidence. But, working here on a small scale, I think is sensible to confine myself in the way in which I have done. Moreover, there are a limited number of theological-ethical arguments and a limited number of methodological options, even if they come in different guises.

PAUL RAMSEY

Paul Ramsey (d. 1988) was the doyen of American Protestant medical ethics over many years. His entry into this sphere of medical-ethical debate came as early as 1956, with an article called 'Freedom and responsibility in medical and sex ethics: a Protestant view', which was a critique of Joseph Fletcher's *Morals and Medicine* of 1954. For present purposes I quote from an article published by Ramsey in 1972.[1] In this article two themes recur. These are the absolute ethical impropriety of embryo experimentation and the accusation that in the new bioethics, the experts are 'playing God'. In the article Ramsey is mainly dealing with IVF.

But the discussion is immediately relevant on both themes to the genetic realm. Ramsey had already developed his thinking on the latter theme in his *Fabricated Man: the Ethics of Genetic Control.*[2] In the present context I confine myself to the charge of bioethics 'playing God'. Thus Ramsey refers to the IVF scientists who must 'mimic nature perfectly' and to 'this artificial mimicry of nature'.[3] Again, 'we can clearly see the extent to which human procreation has already been replaced by the idea of "manufacturing" our progeny' and 'instead, the child as a product of technology is to be brought forth'.[4] I treat Ramsey as standing at one end of a range extending from 'unpermissive' to 'permissive'. Joseph Fletcher represents the latter. Now I explore four reports in the middle part of the range.

'ETHICAL ISSUES IN THE BIOLOGICAL MANIPULATION OF LIFE'

The above phrase is taken from the Report of the World Council of Churches 1979 Conference on 'Faith, Science and the Future' held at MIT, Cambridge, Massachusetts.[5]

MIT followed the activist 1968 Uppsala WCC Assembly, which in turn followed the explosive 1966 Geneva Conference, which introduced revolutionary politics and the critique of technology on to the main WCC agenda – hence the keywords in the title of the Report are 'faith', 'science' and 'unjust'. So a strongly political perspective serves as a presupposition for the ethics of genetic engineering.

> In considering certain biological, genetic and medical manipulations of human beings . . . we must keep sharply in mind the tragic consequences suffered by many people as a result of past eugenic theories and practices. These have all too often been used by groups in power against exploited or disenfranchised minorities[6] . . . [Hence,] the closest scrutiny of the social and economic conditions of . . . [human genetic analysis] application will be continuously required to ensure that such technologies contribute to a just and participatory society.[7]

MIT wants at least to keep in balance both faith and praxis. Thus, theologically, 'as Christians we *believe* that we are both creatures of God and co-creators with him in fulfilling the image He has given us. This *belief* forms our high estimate of human worth.' That *belief* is then immediately linked with *action*. 'We are also guided by the words of Jesus: "Inasmuch as you *do* it unto one of the least of these you *do* it unto me."'[8] *MIT* is the most politically and socially self-conscious of the texts which I examine.

MIT observes a sharp ethical distinction between somatic and germ-line genetic engineering. 'An individual may give informed consent to a change in his own body as in body cell alteration . . . and that change is not transmitted to future generations. *The individual cannot give informed consent to the genetic alteration of subsequent generations.*'[9] Such decisions should be taken in a much broader social context. They cannot be just the products of contracts between scientist and subject.

So, among 'issues of grave concern' is gene replacement. Some features of this can only be ascertained by experimentation, 'first on animals over several generations but eventually on human subjects'. Echoing Ramsey, *MIT* concludes that 'someone has to be the first human experimental subject'.[10]

Critical remarks about theology are not confined to *MIT*. But *MIT* forcefully claims that 'our understanding of the relationship of God to human life and non-human life requires much further development if we are to have a substantial ethical and theological base for evaluating genetic engineering in humans and other vertebrates. In our judgement, *theology does not yet speak in a meaninful way to the problems raised by contemporary biology*'.[11] The meaning of this passage is by no means clear. Partly, it may mean that theology knows how to talk about human beings, but not about non-human, impersonal natural processes.

EXPERIMENTS WITH MAN

Experiments with Man: Report of an Ecumenical Consultation (EWM), edited by Hans-Ruedi Weber, was published in 1969 by the World Council of Churches and Friendship Press, New York. It is divided into three parts – 'A record of convictions and questions', 'Working papers', and 'Meditations on man and his vocation'.

'One aspect of human existence needs close scrutiny now, namely man's power to discover, control, consciously influence and change his environment'.[12] This raises profound questions. Can humans, in fact, influence and change nature and history? What is the purpose of such human dominion over nature and history? What do humans do with their dangerous knowledge? Are there any limits to human power? *EWM* recognizes that there are searching questions to be put *prior* to addressing biotechnological issues.

There are four types of experiment: (1) Routine treatment which, with each patient, is something of an experimental venture. (2) In the course of treatment, data are collected by which existing hypotheses, methods and medicaments are checked and improved and new ones developed. (3) Fundamental clinical research not primarily directed to any immediate therapeutic benefit. But (4) there is biomedical research relating to the manipulation of genetic material. Here *EWM*, from 1969, peers forward and remarks that

> the ethical considerations attending research of this sort would differ markedly from those proper to the other three types. *Here, the investigator and society must somehow act as guarantors of the integrity of future generations by defining the limits permissible in the manipulation of the basic material of the human species.*[13]

'Defining the limits' becomes a recurrent theme in these texts.

EWM is strong on method. It has a useful note on natural law to which I shall refer later. It also goes further than *MIT* in making clear its expectations from theology:

> Today . . . the old theological questions [set in the context and terminology of the humanities, philosophy and social sciences] are [now] raised in the context and terminology of the natural and medical sciences. Much translation and reformulation will be necessary before the rich insights which theology has gained throughout centuries will be of help to biomedical investigators.[14]

EWM explores some theological possibilities along these lines: 'the act of creation is a continuing event; it therefore reveals something not only about the origin, but also about the present history and the *telos* . . . of this universe'. Again, 'can this concept [of natural law be freed] to become a useful theological clue for an ethics of biomedical research which explores the "laws" and mutations of life, nature and body?'. Yet again, 'can the sinfulness of man, and the cross and resurrection of Jesus Christ, which express and answer the deepest predicament and need of man . . . become a clue for the understanding of what biomedical research now discovers about the fragility of life, its destruction, and its effective mechanisms of repair?'.[15]

PERSONAL ORIGINS

I now turn to *Personal Origins* (*PO*).[16] As the subtitle hints, it was produced in the new context created by the setting-up of the Warnock Committee on Human Fertilization and Embryology in 1982. One feature of the working party which deserves some discussion is the discovery of fundamental differences of approach among its members which prevented unanimity. These differences reflected the consequences of the impact of modern science. One approach welcomed the changes and showed a readiness to accept new knowledge. Another saw humans as in danger of losing any serious sense of the boundaries of natural law and of the purposes for which things are created by God. According to that approach we run the risk of overestimating human abilities and our place in relation to nature. But both approaches seek to arrive at judgments about where the moral boundaries for human action on the natural world lie. *PO* then presents 'certain fundamental principles' which set the terms of the discussion. These principles are 'Creation and Natural Law, Order in Nature, Human Responsibility, Eschatology, The Kingdom of God, Jesus Christ and the Divine Purpose'. It is claimed in the text that the six principles provide the boundaries and, indeed, the avenue for proper Christian reflection and judgment. This is an over-optimistic claim. But the six principles (are they really 'principles'?) provide useful headings.

Of the texts on which I am commenting, *PO* appears to have been theologically the most ambitious. That it has not succeeded is probably in part the result of trying to weave into the body of the writing a wide variety of standpoints and attitudes and minority views.

Specifically on genetics, the text seeks to rebut the contention that we are 'the prisoners of our genes'. It claims that the genes we are born with do not necessarily determine our life and health. Even in the limited chemical sense we are more than the sum of our genes.[17]

MANIPULATING LIFE

Manipulating Life[18] differs from all the other reports under consideration with respect to its expressed view that faith has much to learn from the new science. (Is it in this regard more than a coincidence that a large majority of the working party that produced the report were scientists?) For example: 'how can they [sc.

the churches] come to a better understanding of the meaning of faith in the full light of the new knowledge and power that the biological sciences have brought?' 'The creative response to this new knowledge will be one that leads to a deeper understanding of the nature of life and of human responsibility for life.' 'The working out of the relation between faith and this knowledge . . . we have as yet hardly begun'.[19] This view differs greatly from the initial reactions of some churches, which have been of fear, distrust and even calls for prohibition of these new developments.

ML is also different from the other texts in the degree in which it seeks to identify the background causes of the technological ambiguity. Thus: 'of special concern is the reality that the inequitable distribution of power, influence and knowledge increases the possibility of the misuse of technology'.[20] Indeed, 'many of the social and ethical issues that arise in debates about genetic engineering have less to do with genetics than with broader questions of human good'.[21]

As far as somatic and germ-line therapy are concerned, *ML* reiterates, in its call for caution, what has already been noted in, for example, MIT. But again *ML* takes a further and positive step. It argues that 'nonetheless, changes in genes that avoid the occurrence of disease *are not necessarily made illicit merely because those changes also alter the genetic inheritance of future generations*. By overcoming a deleterious gene in future beings, the beneficial effect of such changes may actually be magnified'.[22] Clearly this addition in *ML*, which is not so remarkable when all is said and done, does not allow us to accuse the text of ethical shallowness. For, a few paragraphs further on, the theme is taken up again: 'Many ethical and religious traditions endorse some human freedom to modify or transcend nature. But for us to change *substantially* the germ-line DNA is to directly alter the genetic foundations of the human. In what ways do we, by manipulating our genes in other than simple ways, change ourselves to something less than human?'.[23] However, a factor not mentioned in any of the texts treated here may be relevant, namely whether, on the whole, Christian ethicists are more comfortable dealing with individual rather than with collective or social ethics.[24]

Lastly, it is important to take notice of the high degree of methodological openness which characterises *ML*, in particular the relevance of the *unprecedented* nature of many of the problems about genetics. *ML* does not allow that it is feasible, in these circumstances, wholly to depend on 'precedents from the past to provide answers to questions never asked in the past'.[25]

JOSEPH FLETCHER

Fletcher's *Morals and Medicine* (1954) is commonly regarded as the first Protestant medical ethics in modern times. However, the text which I am using here is Fletcher's article 'Ethical aspects of genetic controls: designed genetic changes in man'.[26] Fletcher is seemingly a polar opposite to Ramsey. He is a consequentialist, a pragmatist, an empiricist, a situationalist, a personalist, and a (sort of) existentialist. The 'definitive question in the ethical analysis of genetic control' is whether we argue deductively *or* from 'empirical data, variable situations and human values to normative decisions'.[27]

Fletcher, as a consequentialist, argues for genetic engineering from, for example, social needs: 'It is entirely possible, given our present increasing pollution of the human gene-pool through uncontrolled sexual reproduction, that we might have to replicate healthy people to compensate for the spread of genetic diseases and to elevate the plus factors available in ordinary reproduction.'[28] As for the morality of the *means* of genetic research, the only objection would be that fertilization or cloning would result directly and instantly in human beings, or in creatures with nascent or proto-human status. Fletcher denies that embryos are such beings.

Ramsey, Cass and others would argue that the laboratory reproduction of human beings is no longer *human* procreation. For Fletcher the crux is the meaning of *human*. He counters as follows:

> Man is a maker and a selecter and a designer, and the more rationally contrived anything is, the more human it is. Any attempt to set up an antinomy between natural and biologic reproduction, on the one hand, and artificial or designed reproduction, on the other, is absurd. The real difference is between accidental or random reproduction and rationally willed or chosen reproduction. In either case it will be biologic – according to the nature of the biologic process'.

Laboratory reproduction is radically human. It is willed, chosen, purposed and controlled – traits which distinguish homo sapiens.[29]

One way of refuting Cass, Ramsey and others of their view is to say that in fact there is, for example, in IVF, very little that is artificial – assisting with the growth and movement of eggs, assisting with the co-location of sperm and egg, and assisting with the journey to the uterus. But Fletcher would want to reduce this (he would say) *high* proportion of biological process if the conception could thereby be the *more* willed, chosen, purposed and controlled. It is because he does not want to sacralize reproductive nature that Fletcher is not disposed to accept that 'in the sequence or progression from aspirin to insulin to artificial kidneys to brain surgery to genetic engineering there is no point at which we can 'change from a clear yes to an absolute no', even though there is a mounting difference in the complexity of the ethical issues posed'.[30]

But the heart of the argument rests upon the definition of the word 'human'. As Richard McCormick observes: 'rational control, therefore, is not the guarantor of human choices but only the condition of their possibility'. It is not clear whether Fletcher has *adequately* considered the 'conflicting interests and values that lie at the heart of ethical problems'.[31]

CONCLUDING COMMENTARY I

The texts exhibit different approaches to natural law. The use in *PO*[32] is virtually unrevised from the Tridentine and post-Tridentine use. Oliver O'Donovan in *Begotten or Made?*[33]: 'we must cherish nature in this place where we encounter it, we must defer to its immanent laws, and we must plan our activities in cooperation with them' is unrevised. In *EWM*, there is a definition on the road to revision: 'man is aware that he must respond to demands which are proper to his realization of himself as a man, demands written into the very fact that he is a man'.[34] Ryan is clear that natural law does not refer to natural processes and laws found by humans looking in on 'nature' from outside. On the contrary, 'only in man, with his capacity to reflect upon and know himself, can there be any question of natural law, the recognition for himself of how he is to act; and only in his case, with his freedom of action, can there be a responsible following or deviation from the "rules" of his nature . . .'. So 'it is left for men to discover for themselves the "rules" built into their nature as human beings, to discover how to use themselves if they are to get the best out of themselves and find their proper human fulfilment. And by natural law is not meant these "rules" as built into them . . . but these "rules" as recognised by men on reflection upon what it is to be human and to find human fulfilment.'[35] This account seems to allow for cultural diversity. Another excellent version for comparison is found in Kelly:

> that [mistaken] approach to natural law would maintain that by finding out how we function *naturally* as human beings (biologically, sexually, genetically, psychologically etc) we discover how God intends us to live. The view I have been describing does not accept this position. It does not believe that 'meaning' is *written* into 'nature' in this way. It believes that 'nature' has to be *read* – and the very reading is an interactive and creative process in which the human mind plays an indispensable interpretative and creative role.[36]

But though it is clearly possible to draw a distinction, as these quotations do, between a less and a more flexible and anthropocentric definition of natural law, the texts offer little or no indication of what the revised doing of natural law would look like in practice.

CONCLUDING COMMENTARY II

The yield of the texts with regard to the purpose and nature of theology is somewhat limited. There are signs of a search for a new theological modality which will be appropriate to genetic and other contemporary issues. The text which is most sensitive to these needs and which tries to say something specific about theological method, namely *PO*, turns out to be somewhat incoherent theologically. The argument in *EWM* concerning the need for 'translation' of past theological concepts is only convincing to a limited degree. This kind of plea often grossly underestimates the problems inherent in the notion of translation. The need in this respect is actually for a dialectical or two-way hermeneutic – a procedure the difficulty of which greatly exceeds popular notions of what is involved in translation. Another possible direction, also suggested in *EWM*, was that 'for this new kind of theological reflection a team is needed in which specialists of biblical, historical and systematic theology work together with people involved in biomedical and other research'.[37] This proposal recognizes the problems posed by specialism of content. It also recognizes the necessity of exposing the different methodologies to each other. The omission of ethics is strange.

CONCLUDING COMMENTARY III

The texts, excepting *MIT* and *ML*, include virtually nothing of the perspectives of the social and political sciences. A terse statement in *ML*, quoted above, tantalizingly points in this direction. 'Of special concern is the reality that the inequitable distribution of power, influence and knowledge increases the possibility of the misuse of technology'.[38] In fact, deductivist theology finds it very difficult to engage with these socio-political dimensions. Of the texts analysed, Fletcher's seems the one which, with modification, promises the most. In particular, the 'situation' to which this 'situationalism' refers has to be taken in a much broader sense than Fletcher allows. But there is nothing about Fletcher's method which precludes the scope of reflection being enhanced by further genres of theological and ethical consideration.

In her useful philosophical reflection on the notion of 'playing God', Ruth Chadwick isolates two types of context. In the second, where the objection to playing God functions as a counsel, it does not seem to be strong enough to rule out particular courses of action. 'It cannot provide us with any definite boundaries which should not be crossed.'[39] I am inclined to say the same of theology and of theological ethics. 'This is [rather] a matter for consequentialist assessment.'

The only tangible ethical dispute raised at any length by the texts concerns the questions about somatic-cell gene therapy and germ-line therapy. Anderson stated in 1987 that 'essentially all observers have stated that they believe that it would be ethical to insert genetic material into a human being for the sole purpose of medically correcting a severe genetic defect in that patient'.[40] Somatic-cell gene therapy is ethically acceptable 'if carried out under the same strict criteria that cover other new and experimental medical procedures'.[41] These requirements 'simply state that the new treatment should get to the area of disease, correct it, and do more good than harm'.[42]

By contrast, therapy of germ-line cells 'would require a major advance in our present state of knowledge'. The technique has 'a high failure rate', can produce a 'deleterious result', and would have 'limited usefulness'.[43] But, even when the technical problems are ironed out, are there other questions, such as whether 'the transmitted gene itself, or any side effects caused by its presence, [will] adversely affect the immediate offspring or their descendants'.[44] In fact Anderson concludes that, given adequate experience with somatic therapy, adequate animal studies and public awareness and approval, germ-line therapy would be 'ethical and appropriate'.[45]

All the criteria put forward by Anderson are medical-technical, *except* that concerning 'public awareness and approval'.[46] This reference to public awareness and approval is too weak and too general a way to describe those many, many non-medical areas of ethical perplexity. The latter come to expression very clearly in the Report of the Enquete Commission on *Opportunities and Risks in Genetic Technology*. For example, the Commission took the view that human germ-line therapy 'should be forbidden by law, because, among other things, it might open the way for selective breeding of human beings'.[47] It may be argued that such a consideration is of a universal kind and should apply in all places and at all times. Or, on the other hand, it could be claimed that for the German nation *not* to proscribe in this way would be an immoral symbolic rejection of the memory of the Nazi death camps and of the Hitler doctors. This was the opening consideration of *MIT*, discussed above.

But there is a counter-argument. So important for the genetic future of (parts of) the human race might germ-line therapy be, that the risk of misuse must be accepted whilst, by the most careful legislation and regulation, everything possible was done to reduce to a minimum that risk. The impression is gained from some of the texts analysed above that there was a sigh of ethical relief that, apparently, there exists a clear-cut choice between somatic and germ-line therapy. But if, as I should contend, the choice is in fact much more complicated, so the range, character and use of the theological and other resources would in turn be all the more complicated. That, however, is to take the ethical challenge *more* seriously, not to undermine it.

CONCLUDING COMMENTARY IV

I have reserved quite sharp criticism for much of the theological prescription that appears in the texts which I have analysed. This does not necessarily render the texts valueless. It shows, for example, that theology is not a poor kind of science; nor is it an untidy kind of philosophy. On the other hand, theology is not in business to recover and dogmatically to repeat earlier constellations of thought. Theology is better understood as the bringing to conscious form of the varied and sometimes inconsistent life-experience of a particular historical-ethical-religious stream or movement, so that this experience can be tested, explored, developed, communicated and shared. The peculiar task is, using the jargon, to combine the experiences of 'historical existence' and 'transcendence' in the area of genetic engineering. It turns out to be a tough assignment. What *that* has to say about 'historical existence' and about 'transcendence' is another story.

NOTES

1 Paul Ramsey, 'Shall we "reproduce"?', *Journal of the American Medical Association* 220 (5 June 1972), 1346–50, 1480–5.
2 Paul Ramsey, *Fabricated Man: The Ethics of Genetic Control* (New Haven: Yale University Press, 1970).
3 Ramsey, 'Shall we "reproduce"?', 1347.
4 ibid., 1480f.
5 *Faith and Science in an Unjust World*, ed. Paul Abrecht, vol. 2 (Geneva: World Council of Churches, 1980); henceforth *MIT*.
6 *MIT* 49.
7 ibid.
8 ibid., 49f.; my italics.
9 ibid., 54; my italics.
10 ibid.
11 ibid., 55f.
12 *EWM*, 7.
13 ibid., 13.
14 ibid., 18.
15 ibid., 19f.
16 This text is subtitled *The Report of a Working Party on Human Fertilisation and Embryology of the Board for Social Responsibility [of the General Synod of the Church of England]*. It appeared from CIO Publishing, London, 1985: henceforth *PO*.
17 *PO*, 12f.
18 *Manipulating Life: Ethical Issues in Genetic Engineering* (Geneva: World Council of Churches, 1982), 1; henceforth *ML*.
19 *ML*, 3.
20 ibid., 3.
21 ibid., 5.
22 ibid., 7.
23 ibid., 8.
24 See, for example, K. Stendahl, 'St Paul and the introspective conscience of the West', *Harvard Theological Review* 56 (1963), 119–215.
25 *ML*, 11.
26 *New England Journal of Medicine* 285 (1971), 776–83.
27 ibid., 777.

28 ibid., 779.
29 ibid., 780.
30 ibid., 782.
31 Bernard D. Davies in Richard A. McCormick, 'Genetic medicine: notes on the moral literature', *Theological Studies* 33 (1972), 535.
32 *PO*, 18.
33 Oliver O'Donovan, *Begotten or Made?* (Oxford: Clarendon Press, 1984), 5.
34 *EWM*, 20f.
35 Columba Ryan, 'The traditional concept of natural law: an interpretation', in Illtud Evans (ed.), *Light on the Natural Law* (London: Burns & Oates, 1965), 22f.
36 Kevin T. Kelly, 'The Embryo Research Bill: some underlying ethical issues', *The Month* 23 (1990), 119.
37 *EWM*, 19.
38 *ML*, 3.
39 Ruth F. Chadwick, 'Playing God', *Cogito* 3 (autumn 1989), 192.
40 W. F. Anderson, 'Human gene therapy: scientific and ethical considerations', in Ruth F. Chadwick (ed.), *Ethics, Reproduction and Genetic Control* (London: Croom Helm, 1987; London and New York: Routledge, 1990), 149.
41 ibid.
42 ibid. 150
43 ibid. 155ff.
44 ibid. 157.
45 ibid. 159.
46 'Linda Bullard claims that genetic engineering is "*inherently* eugenic in that it always requires someone to decide what is a good and a bad gene"', in Helen Bequaert Holmes and Laura M. Purdy (eds), *Feminist Perspectives in Medical Ethics* (Bloomington and Indianapolis: Indiana University Press, 1992), 193.
47 W. M. Katenhausen, 'Prospects and risks of genetic engineering', in David Weatherall and Julian H. Shelley (eds), *Social Consequences of Genetic Engineering* (Amsterdam, New York and Oxford: Excerpta Medica, 1989), 181.

INDEX

The words 'biotechnology', 'gene', 'genetic engineering', 'somatic', and 'cell-line' are used so frequently in the text that they are not included in the Index.